T0419369

ENVIRONMENTAL HEALTH - PHYSICAL, CHEMICAL AND BIOLOGICAL FACTORS

NEW DEVELOPMENTS IN BIODIVERSITY CONSERVATION

Environmental Health - Physical, Chemical and Biological Factors

Additional books in this series can be found on Nova's website under the Series tab.

Additional E-books in this series can be found on Nova's website under the E-books tab.

Environmental Science, Engineering and Technology

Additional books in this series can be found on Nova's website under the Series tab.

Additional E-books in this series can be found on Nova's website under the E-books tab.

ENVIRONMENTAL HEALTH - PHYSICAL, CHEMICAL AND BIOLOGICAL FACTORS

NEW DEVELOPMENTS IN BIODIVERSITY CONSERVATION

THOMAS W. PACE
EDITOR

Nova Science Publishers, Inc.
New York

For permission to use material from this book please contact us:
Telephone 631-231-7269; Fax 631-231-8175
Web Site: http://www.novapublishers.com

Library of Congress Cataloging-in-Publication Data

New developments in biodiversity conservation / editor, Thomas W. Pace.
p. cm.
Includes bibliographical references and index.
ISBN 978-1-61324-374-9 (hardcover : alk. paper) 1. Biodiversity conservation. I. Pace, Thomas W. II. Title.

QH75.N44 2011
639.9--dc23

2011012561

Published by Nova Science Publishers, Inc. ✝ New York

CONTENTS

PREFACE

This book presents current research in the study of biodiversity conservation. Topics discussed include agricultural land-use in forest frontier areas; utilization of bovids in traditional folk medicine and their implications for conservation; conservation and management of the biodiversity in a hotspot; surface stratification of soil nutrients in no till limits nutrient availability; forage resources renewal and biodiversity conservation and outcomes of invasive plant-native plant interactions in North American freshwater wetlands.

Chapter 1 – In this paper author develop a spatially explicit economic land-use model that gives insights into the determinants of land-use patterns and how these patterns are affected by policy changes. The model explicitly takes into account the decision-making process as to why and where farmers convert the use of forest land. This is different from previous spatially disaggregated models – such as simulation models – where the underlying decision-making process is imposed. The micro-economic focus in this paper is crucial for understanding the ongoing human-induced land-use change process and is essential in the land-use change literature – that is dominated by natural scientists focusing on geophysical and agro-climatic processes. The author's model is extremely valuable to inform land-use policy as it specifies how individual decision makers will react to policy and other exogenous changes in their environment and how this response will alter the landscape. The model is derived from the von Thunen-Ricardo land rent model that describes land-use patterns as a result of variability in geophysical land attributes and differences in location and transport costs. However, this model is valid only under certain assumptions and is less suited to describe land-use patterns in forest frontier areas characterized by semi-subsistence agriculture and imperfect markets. Author refine the model to account for the fact that agricultural prices and wages might be endogenously determined and households cannot be considered as profit maximizing agents. Author empirically estimate the model for a forest-frontier area in Indonesia using a combination of data from satellite image interpretation, GIS data and a socio-economic survey data. The results demonstrate that differences in Ricardian land rent are important in determining spatial land-use patterns. However, author do not find evidence in support of the von Thunen idea that land-use patterns are determined by differences in transport costs. Rather the labor intensity of land-use systems, population levels, the access to technology and household characteristics matter. This has important implications for forest conservation and land-use policy. In addition, the refinement of the von Thunen-Ricardo land rent model – which incorporates more realistic descriptions of economic behavior – is justified by the empirical results.

Chapter 2 – Animals and products derived from different organs of their bodies have constituted part of the inventory of medicinal substances used in various cultures since ancient times. Regrettably, wild populations of numerous species are overexploited around the globe, the demand created by the traditional medicine being one of the causes of the overexploitation. Mammals are among the animal species most frequently used in traditional folk medicine and many species of bovids are used as medicines in the world. The present work provides an overview of the global usage of bovids in traditional folk medicine around the world and their implications for conservation. The results demonstrate that at least 55 bovids are used in traditional folk medicine around the world. Most of species (n=49) recorded were harvested directly from the wild, and only six species of domestic animals. Of the bovids recorded, 50 are included on the IUCN Red List of Threatened Species and 54 are listed in the CITES. By highlighting the role played by animal-based remedies in the traditional medicines, author hope to increase awareness about zootherapeutic practices, particularly in the context of wildlife conservation.

Chapter 3 – New Caledonia is a peculiar hotspot, a small-sized island (ca. 17,000 km^2), relatively isolated from any continent (ca. 1200 km from Australia), with moderately high mountains and complex orography. Its biota is very rich in endemic species and highly endangered. The author's analysis of the number of references in systematics, ecology and conservation shows that its biota attracts attention of scientists since long ago. In a first and long period, references focused on the description of the biodiversity. More recently these descriptions were intensified and complemented by ecological and later by conservation studies. Recent researches on phylogenetics and biogeography indicated that the biota of New Caledonia is characterized by short range endemism and rarity, with three patterns of endemism: (I) species regionally endemic to New Caledonia distributed in the whole range of an ecosystem; (II) short range endemics with parapatric/allopatric distributions; (II) short range endemics with disjunct distributions. Researches on conservation showed that this biota is highly endangered due to three main threats: fire, mining and invasive species. In this chapter author detail these threats and elaborate a model based on their frequency and spatial distribution to understand how they could affect species with contrasting patterns of endemicity. The author's analysis show that the conservation of the biodiversity in a context where species are dominantly short range endemics and rare is a main problem to be faced by New Caledonian authorities as well as by scientific researchers that must provide the basis for political decisions. Regardless the biogeographical pattern of endemism the chances of survival of rare species with short ranges in the case of large scale habitat destruction are quite low. In the case of threats that are more restricted in area, the loss of a species disjunctly distributed is more problematic in terms of loss of phylogenetic diversity. In this case, speciation by niche conservatism can be hypothesized to be less frequent, thus each species can be implied to be more original and in stronger need of conservation by itself. In addition, due to the distances from one another, the number of closely related species in a small island can be much lower than in the case of species with parapatric/allopatric distribution. Fires, mining and introduced species need special control. The two firsts for the habitat destruction they promote over extensive areas and the later by the possibility of continuing to endanger even in areas officially protected.

Chapter 4 – Research has shown that the less mobile elements in the soil, such as P, K and Zn, become stratified in the surface 50 mm of soil after several years of continuous no-till cropping. This causes nutritional constraints to productivity when the surface soil becomes dry unless the sub-surface soil is highly fertile or receives appropriate levels of fertiliser. No-till crops grown in the semi-arid, subtropical environment of central Queensland, Australia, are particularly prone to nutritional disorders as a result of surface stratification. The surface soil can remain dry for long periods during crop growth while the roots obtain water stored in the heavy clay soils during the preceding fallow. Consequently nutrient deficiency symptoms have appeared despite apparently satisfactory levels of those same nutrient elements in the 0-100 mm layer, which is the standard sampling depth used to assess soil fertility. The original aim of the 13-year experiment reported here was to assess the effect of tillage frequency and intensity, and stubble retention and removal, on soil water storage, soil nutrient status, and the growth and yield of rainfed grain sorghum in central Queensland. During the first 7 years of research, the mean yields of no till and traditional tillage (disc plough and scarifier) without fertiliser application were not significantly different. As the experiment continued, it became apparent that supplementary nutrition was required. The placement of appropriate levels of fertiliser elements (P, K and Zn) 100 mm deep and 50 mm to the side of each row resulted in large yield responses to no till. The yield of no till in comparison with traditional tillage (both with stubble retained and appropriate fertiliser application) was 147%, 129% and 174% in the final three crops, respectively. These responses also reflect the outstanding potential of longer-term no till that has been demonstrated in other research work in central Queensland. Since mixing of the surface soil (by tillage) gave similar responses, it was concluded that some crop nutrients had become concentrated in the dry surface soil and were therefore inaccessible to plant roots during dry periods.

Chapter 5 – Metallothioneins (MTs) are ubiquitous and inducible proteins characterized by low molecular mass (Mr 6-8 kDa), high Cys content (20-30%) but no aromatic or His residues, and strong affinity to binding toxic metals (Cd, Hg, Ag, Pb) in metal-thiolate clusters. Due to their induction by a variety of stimuli, MTs are considered suitable biomarkers in the medical and environmental fields. The protective role of MTs from Cd toxicity and lethality is well-established. Although MT assessment is a difficult task, the accurate measurement of MT is mandatory in order to assess its biomarker potential and to identify new outstanding biological roles. Author have developed a highly specific, sensitive, and reliable method for total MT quantification in unheated extracts by reversed-phase high-performance liquid chromatography coupled to fluorescence detection (RP-HPLC-FD). A derivatization protocol with monobromobinane, a thiol-specific fluorogenic reagent, is required after heat-, SDS- EDTA- and DTT-treatment. SDS-polyacrylamide gel electrophoresis was used to confirm the identity of the mBBr-labeled MT peak resolved by RP-HPLC-FD. The method has been successfully used to quantify MT content in the digestive gland of various clam species from Southern Spanish sites with different metal levels, and also in the liver of fish injected with different Cd, Cu and Hg doses. MT levels obtained by RP-HPLC-FD in non-heated extracts were significantly higher when compared to those obtained by other well-established assays relying on solvent precipitation (spectrophotometry) or heating (differential pulse polarography) pre-purification steps.

Chapter 6 – After several decades of marginalization within farming systems, European rangelands are now being challenged to contribute to the conservation of ecological habitats and biodiversity. One of the main challenges, supported by the European Union incentives, relates to the reconciliation of livestock farmers' grazing practices and to the control of dominant plant dynamics, especially those of shrub species, which includes maintaining them at density levels appropriate for both habitat conservation and forage resources production. In this chapter, author aim to identify reasons for the difficulty in designing relevant management practices, with focus on the interlinkage of knowledge produced by animal sciences and plant population ecology. From the point of view of these two disciplines, author stress the importance of taking into account the reciprocal interactions between ruminants' foraging strategy and shrubs' demographic behavior. A series of results is given for the author's experiments on rangelands encroached by Scotch Broom shrubs (Cytisus scoparius L.Linck) and grazed by ewes. Considering the dynamics of ewes' behavioral patterns, author argue for a description of heterogeneous vegetation that recognizes feed items and their functionality for ruminants in maintaining their intake level in a fenced paddock. And considering the dynamic response of Scotch Broom shrubs to browsing, author also argue for demographic models that would recognize the importance of both the selective offtake of plant organs and the new browsing-induced demographic behavior of shrubs. These results enabled to identify the plant community as a mandatory intermediate object, and the plant organs as a key organization level at which these two processes interact. Author propose an original conceptual framework that interlinks the two processes and recognizes the specific organizational levels and time frames. This framework should facilitate the identification of prospective research issues such as the differential impact of browsing on shrub demography according to plant organs and life stages consumed, or the effect of different vegetation states of a plant community on selective browsing among shrub organs. For rangeland management, the framework brings out the importance of greater precision in identifying the targets, and in particular the target plant organs and target life stages in shrub demography control. Considering this objective the choice of the season for grazing a given fenced pasture should also be made bearing in mind the global feeding offer within the plant community.

Chapter 7 – Freshwater tidal wetlands are productive and often support high biodiversity. While there have been some quantitative studies of the effects of many invasive plants in North American freshwater wetlands, much is still assumed for a number of invasive species. The outcomes of invasive-native plant interactions, and the factors involved, are more subtle than assumed. author reviewed the statistical and anecdotal results of published studies on the impacts of invasive plants with large potential ranges in freshwater tidal wetlands of North America. A number of studies reported little or no change in native species richness and diversity, and found outcomes differed depending upon several specific factors. author recommend that more empirical research be conducted on the interactions of these species (both in field and greenhouse) with native plants in freshwater tidal wetlands because very little information is available and yet many management decisions have already been made.

In: New Developments in Biodiversity ... ISBN: 978-1-61324-374-9
Editor: Thomas W. Pace

Chapter 1

AGRICULTURAL LAND-USE IN FOREST FRONTIER AREAS: THEORY AND EVIDENCE FROM INDONESIA

Miet Maertens[1], **Manfred Zeller**[2] **and Regina Birner**[3]*
[1]Agricultural and Food Economics Section, Department of Earth and Environmental Sciences, Katholieke Universiteit Leuven, Belgium
[2]Institute for Agricultural Economics and Social Sciences in the Tropics and Subtropics, Universität Hohenheim, Germany
[3]International Food Policy Research Institute, Washington D.C., U.S.A.

ABSTRACT

In this paper we develop a spatially explicit economic land-use model that gives insights into the determinants of land-use patterns and how these patterns are affected by policy changes. The model explicitly takes into account the decision-making process as to why and where farmers convert the use of forest land. This is different from previous spatially disaggregated models – such as simulation models – where the underlying decision-making process is imposed. The micro-economic focus in this paper is crucial for understanding the ongoing human-induced land-use change process and is essential in the land-use change literature – that is dominated by natural scientists focusing on geophysical

* Corresponding author: Miet.Maertens@ees.kuleuven.be

and agro-climatic processes. Our model is extremely valuable to inform land-use policy as it specifies how individual decision makers will react to policy and other exogenous changes in their environment and how this response will alter the landscape.

The model is derived from the von Thunen-Ricardo land rent model that describes land-use patterns as a result of variability in geophysical land attributes and differences in location and transport costs. However, this model is valid only under certain assumptions and is less suited to describe land-use patterns in forest frontier areas characterized by semi-subsistence agriculture and imperfect markets. We refine the model to account for the fact that agricultural prices and wages might be endogenously determined and households cannot be considered as profit maximizing agents.

We empirically estimate the model for a forest-frontier area in Indonesia using a combination of data from satellite image interpretation, GIS data and a socio-economic survey data. The results demonstrate that differences in Ricardian land rent are important in determining spatial land-use patterns. However, we do not find evidence in support of the von Thunen idea that land-use patterns are determined by differences in transport costs. Rather the labor intensity of land-use systems, population levels, the access to technology and household characteristics matter. This has important implications for forest conservation and land-use policy. In addition, the refinement of the von Thunen-Ricardo land rent model – which incorporates more realistic descriptions of economic behavior – is justified by the empirical results.

1. INTRODUCTION

There are two distinct aspects related to land use: the quantity or the rate of land-use change and the location of these changes. From an environmental viewpoint the location of land-use change is as important as its magnitude (Nelson and Geoghegan, 2002). For example, for biodiversity protection it is not only important to know the physical extent of deforestation but also the degree to which it affects critical habitats in specific locations. Also for policy makers it is important to anticipate where land-use changes are likely to take place. For example, road development might trigger land-use changes and the exact location of new roads might have a large impact on the landscape.

In this chapter we develop a spatially explicit economic land-use model that gives insights into the determinants of land-use patterns and how these patterns are affected by policy changes. The focus is on a forest frontier area where agricultural settlers are the main agents in land use conversion. The model explicitly takes into account the decision-making process as to why and where smallholder farmers convert the use of forest land; which is crucial for understanding the ongoing human-induced land-use change process and is essential in the land-use change literature – that is dominated by natural scientists focusing on geophysical and agro-climatic processes. Our model is extremely valuable to inform land-use policy as it specifies how individual decision makers will react to policy and other exogenous changes in their environment and how this response will alter the landscape. We discuss the theoretically derived implications of the model and empirically estimate the model for a forest-frontier area in Indonesia using a combination of data from satellite image interpretation, GIS data and a socio-economic survey data.

The chapter is structured as follows. In the next section we give a brief overview of spatially explicit land-use models. In section three we develop a land rent model that can explain forest conversion and agricultural land expansion in forest frontier areas. Section four deals with the empirical estimation of the derived model for a forest frontier area in Indonesia. In a final section we draw some conclusions.

2. Spatial Land Use Models

Location matters and questions concerning spatial patterns of land use and the location of land use change have been addressed in spatially disaggregated land use models. For a long time geographers and natural scientists have dominated spatially explicit land-use analysis with statistical and simulation models (Irwin and Geoghegan, 2001). Some of these studies have focused solely on the geographic and biophysical explanation of spatial land-use patterns (e.g. Liu *et al.*, 1993; Gobin and Feyen, 2002). In such models the role of geophysical factors such as soil type, topography, etc. and landscape elements such as distance to roads, patchiness of different land-use types, etc. is emphasized. Other geographic studies did include socioeconomic variables such as population, wealth and technology in their spatial statistical land-use analysis (e.g. Fox *et al.*, 1994; Mertens and Lambin, 1997; Serneels and Lambin, 2001). However, the selection of socioeconomic variables is often *ad hoc* and the results do not give insights into the decision-making process. Also spatial simulation models that consider socioeconomic factors are not able to model economic responses because the underlying decision-making process is imposed (e.g. Verburg *et al.*, 1999; Verburg *et al.*, 2000; Verburg and Veldkamp, 2001). Geographers and natural scientists have modeled land use in a spatially explicit way, however with less emphasis on understanding the underlying economic processes that lead to spatial land-use patterns.

More recently, questions concerning the location of land use, have also attracted the attention of agricultural economists. Chomitz and Gray (1996) took the lead in describing the landscape using an economic model. Spatially explicit economic land-use models initially focused on the role of roads and geophysical land attributes in affecting the landscape through influencing households' economic behavior (Chomitz and Gray, 1996; Nelson and Hellerstein, 1997). Later, also the impact of population growth and the role of protected areas were addressed (Cropper *et al.*, 2001; Deininger and Minten, 2002; Müller and Zeller, 2002). Access to credit, tenure security (Deininger and Minten, 2002) and the level of technology (Müller and Zeller, 2002; Vance and Geoghegan, 2002) were also taken into account. Vance and Geoghegan (2002) also included household demographic composition in a spatial land-use model. The strength of spatial economic land-use models is that they give insights into the decision-making process as to where people convert the use of land. They are very well suited to model how individual decision makers will react to policy and other exogenous changes and how this response will alter the landscape.

3. A Spatial Economic Land Use Model

The von Thunen – Ricardo Land Rent Model

Spatially explicit economic land-use studies are generally based on an analytical model derived from the land rent theory of von Thunen and Ricardo. An analytical approach was elaborated by Chomitz and Gray (1996) and applied by Nelson and Hellerstein (1997); Cropper *et al.* (2001); Deininger and Minten (2002); Munroe *et al.* (2002) and Müller and Zeller (2002).

A plot of land is assumed to be allocated to the use that brings in the highest rent. Following von Thunen, land rent decreases with distance to a central market because of increasing transport costs, which result in lower output prices and higher input prices (von Thunen, 1826). Since transport costs differ between crops, the land-use pattern will exist of concentric rings of different land-use types around a central market (figure 1). High value crops that are more difficult to transport (such as vegetables) will be located closer to urban centers than more bulky crops with a lower value and lower transport costs (such as grains). The original von Thunen model assumes a featureless plain surrounding a central market (von Thunen, 1826). This model was refined by Chomitz and Gray (1996) and Nelson and Hellerstein (1997) to emphasize the importance of roads. Transport costs do not only increase with Euclidian distance to the market but also with the difficulty of accessing the market.

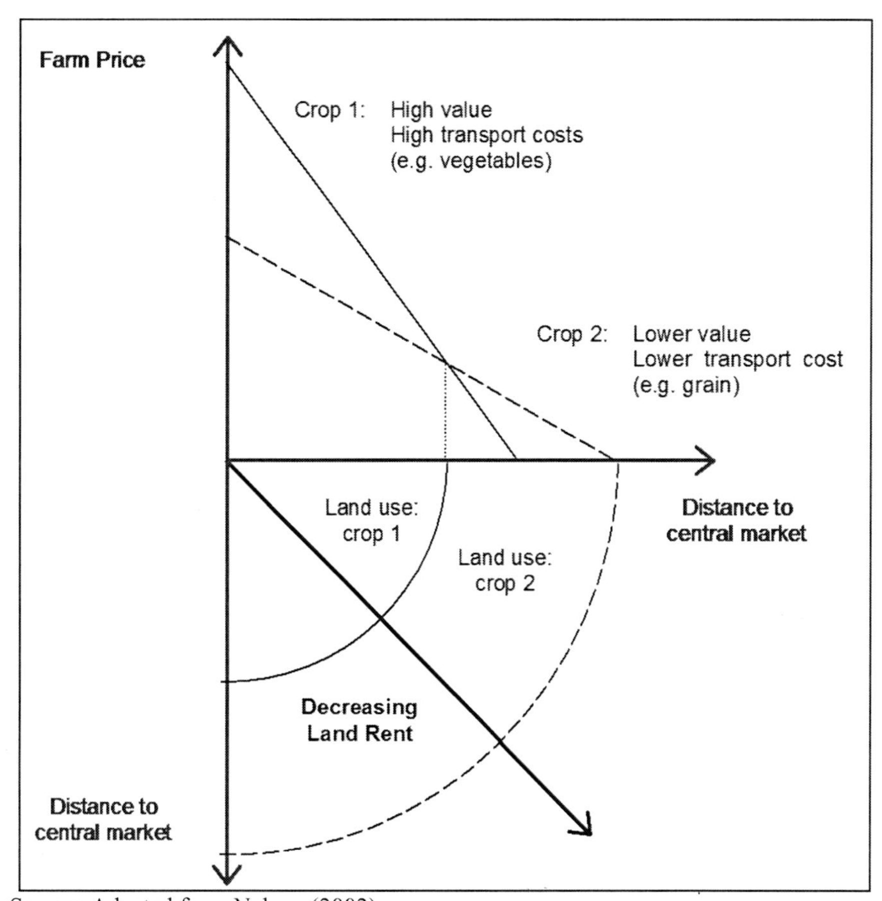

Source: Adapted from Nelson (2002).

Figure 1. The von Thunen Land Rent – Land Use Model.

In addition to the relative location of markets also the location with respect to roads determines land rent.

The model was further elaborated by adding land-rent features according to the work of Ricardo. Next to access to markets and roads also natural land attributes matter. Differences in geophysical characteristics bring about variability in natural land productivity. This determines (potential) agricultural yields and therefore influences land rent. By adding roads and Ricardian features, the original von Thunen model has become more complex. However, the basic insight of the importance of location and transport costs in determining land rent and land use remains (Nelson, 2002).

Shortcomings of the von Thunen Ricardo Land Rent Model

The von Thunen - Ricardo land rent model is valid only under certain assumptions (Lambin *et al.*, 2000; Nelson and Geoghegan, 2002). First, it is assumed that a plot of land will be converted if it is profitable to do so. So, perfect markets and profit-maximizing behavior are implicitly assumed (Nelson and Geoghegan, 2002). This assumption is too simplified for the case of rural areas in developing countries. In such areas markets are often highly imperfect or even missing, which makes households' production decisions to be interlinked with consumption decisions? With missing rural labor markets it is more realistic to assume that households trade-off the utility of farm income and consumption with the disutility of labor rather than maximizing farm profits.

Second, the model assumes that there is one central market where all outputs and inputs, including labor are traded. This might resemble peri-urban areas or isolated commercial settlements reasonably well. However, the land-use pattern in forest frontier areas, characterized by smallholder and semi-subsistence agriculture, is likely to be more diffused and not well described by the von Thunen model (Mertens and Lambin, 1997; Lambin *et al.*, 2000).

Third, it is assumed that agricultural prices in the central market are not influenced by changing land use and product supply. One should be aware that this does not at all hold for products with an inelastic demand such as basic food crops. The assumption is more reasonable in the case of export crops with prices determined by world market conditions and not much affected by local supply.

Fourth, spatial differences in prices are solely related to differences in transport costs. The assumption of increasing input costs with distance to markets is very plausible for bulk inputs such as fertilizers but is less clear for labor inputs (Chomitz and Gray, 1996). Wages do not necessarily increase with distance to the market. On the one hand, opportunity costs of labor might be higher closer to markets because of more profitable off-farm employment opportunities. On the other hand, labor might be scarcer and hence more expensive in less densely populated areas further away from urban centers and markets.

In the next section we elaborate on the von Thunen - Ricardo land rent model by adding some new features that allow us to relax some of these assumptions.

A Refinement of the von Thunen - Ricardo Land Rent Model

As in the spatial land-use model developed by Chomitz and Gray (1996) we assume there is a potential rent attached to each possible land use (k) of each plot of land (i). Each plot will be allocated to the use with the highest rent (R_{ik}). The land rent R_{ik} is the difference between the value of outputs and inputs for land use k at plot i. We assume that the output for each land use k follows a Cobb Douglas production function with capital (k_k), labor (l_k) and land (a_k) inputs, a productivity shifter (S_k) and constant marginal returns to scale.

$$q_k = S_k k_k^{\alpha_k} l_k^{\beta_k} a_k^{\gamma_k} \text{ with } \alpha_k, \beta_k, \gamma_k > 0$$
$$\alpha_k + \beta_k + \gamma_k = 1 \tag{1}$$

The output per unit of land (Q_{ik}) for land use k at plot i is then a function of the capital (K_{ik}) and labor (L_{ik}) inputs per unit of land and the plot-specific productivity shifter (S_{ik}).

$$Q_{ik} = S_{ik} K_{ik}^{\alpha_k} L_{ik}^{\beta_k} \text{ with } \alpha_k, \beta_k > 0$$
$$0 < \alpha_k + \beta_k < 1 \tag{2}$$

The productivity shifter includes the available technology (S_{1v}) and the natural productivity of the plot (S_{2i}) and can be expressed as the product of these factors:

$$S_{ik} = \lambda_{ok} S_{1v}^{\lambda_{1k}} S_{2i}^{\lambda_{2k}} \tag{3}$$

The available technology does not vary for single plots but is assumed to be specific to the village (v) in which the plot is located. Natural productivity is plot specific and includes agro-climatic and geophysical aspects that determine the suitability of the plot for cultivation: slope, soil type, climate etc. The way in which the available technology (S_{1v}) and natural suitability (S_{2i}) shift productivity is specific for each land-use type k. Besides the labor inputs on the plot (L_{ik}), we assume there is an additional labor cost related to an increased walking time to distant plots. This costs is plot-specific and increases exponentially with the distance (D_i) to the village. We specify output prices for different crops (P_k), the price of capital inputs (C), wages (W) and express the potential land rent associated with allocating plot i to land use k as follows:

$$R_{ik} = P_k Q_{ik} - CK_{ik} - WL_{ik} \exp(\gamma D_i) \tag{4}$$

Substituting (2) and (3) in (4) we can derive the optimal amounts of labor and capital inputs for each land use k at each plot i:

$$L_{ik} = \left[\alpha_k^{\alpha_k} \beta_k^{1-\alpha_k} \lambda_{0k} S_{1v}^{\lambda_{1k}} S_{2i}^{\lambda_{2k}} \left(\frac{1}{\exp(\gamma D_i)} \right)^{1-\alpha_k} \left(\frac{P_k}{C} \right)^{\alpha_k} \left(\frac{P_k}{W} \right)^{1-\alpha_k} \right]^{1/1-\alpha_k-\beta_k} \tag{5}$$

$$K_{ik} = \left[\alpha_k^{1-\beta_k} \beta_k^{\beta_k} \lambda_{0k} S_{1v}^{\lambda_{1k}} S_{2i}^{\lambda_{2k}} \left(\frac{1}{\exp(\gamma D_i)} \right)^{\beta_k} \left(\frac{P_k}{C} \right)^{1-\beta_k} \left(\frac{P_k}{W} \right)^{\beta_k} \right]^{1/1-\alpha_k-\beta_k} \tag{6}$$

Substituting (2), (3), (5) and (6) in (4) and rearranging, results in expressing the land rent for plot i and land use k as:

$$R_{ik} = P_k \left[\alpha_k^{\alpha_k} \beta_k^{\beta_k} \lambda_{0k} S_{1v}^{\lambda_{1k}} S_{2i}^{\lambda_{2k}} \left(\frac{1}{\exp(\gamma D_i)} \right)^{\beta_k} \left(\frac{P_k}{C} \right)^{\alpha_k} \left(\frac{P_k}{W} \right)^{\beta_k} \right]^{1/1-\alpha_k-\beta_k} \tag{7}$$

Farm-gate output and input prices and local wages are endogenous and unobserved. We assume that households are integrated in agricultural output and input markets, and that differences in local prices are determined by transport costs from the village to the market. In equation (8) we specify that the price of agricultural outputs relative to capital inputs decreases with distance or costs-of-access from the village to the central market (Z_v). The factor δ_{1k} represents transport costs, which are crop specific. The factor is negative indicating that local output prices decrease with distance to the market and input prices increase. The factor δ_{ok} represents the price of agricultural outputs relative to capital inputs in the central market.

$$\frac{P_k}{C} = \exp(\delta_{ok} + \delta_{1k} Z_v) \tag{8}$$

Labor markets might be missing or highly imperfect due to immobility of labor in the short run and lack of off-farm employment opportunities in remote rural areas. We assume that labor is exchanged locally and express village-level relative wages as a function of the village population (equation 8). A higher population (A_v) increases the supply of labor, increases the demand for agricultural products and therefore increases the price of agricultural outputs relative to wages. The specification depends on the type of crop (γ_{3k}). The price of basic food crops is likely to change much more in response to population growth while the price of export crops might be unaffected. The specification additionally depends on village-specific characteristics (γ_{1v}) and characteristics specific to the village population (γ_{2a}). The way population growth affects labor supply and food demand strongly depends on the age composition and the share of workers versus dependents in the population. Also, the functioning of markets matters. If household are integrated in off-farm labor markets, the effect of population growth on relative wages might be weak. In addition, other unobserved village and household characteristics might influence village-level relative prices.

$$\frac{P_k}{W} = \gamma_{1v}\gamma_{2a}A_v^{\gamma_{3k}} \tag{9}$$

Substituting the expressions (8) and (9) for relative prices in (7), taking natural logarithms, combining all constant terms and combining unobserved effects in an error term we obtain:

$$\ln R_{ik} = a_{ok} + a_{1k}\ln S_{1v} + a_{2k}\ln S_{2i} + a_{3k}D_i + a_{4k}Z_v + a_{5k}\ln A_v + \varepsilon_{ik} \tag{10}$$

The model presented here differs in some aspects from the model elaborated by Chomitz and Gray (1996). Instead of considering plot-specific absolute prices for agricultural outputs and inputs, we considered relative prices at the village level. With this specification we are able to not solely attribute differences in local prices to differences in transport costs to a major market. We account for the endogeneity of agricultural prices and wages by assuming local level relative prices to depend on household- and village-specific characteristics. In addition, our model is not necessarily based on the assumption of perfect markets and profit-maximizing behavior. The model can be put in a framework of utility maximization by assuming differences in relative (implicit) wages to be household specific and ascribing them to differences in the marginal rate of substitution of income (or consumption) for leisure.

Implications of the Model

First, equation (10) indicates that the land rent for agricultural land use increases with improved agricultural technologies (S_{1v}). Second, favorable natural characteristics (S_{2i}) leading to a higher natural productivity also increase the land rent for agricultural land use. Third, the coefficients a_{3k} and a_{4k} are negative and the land rent diminishes as the distance from the plot to the village (D_i) and from the village to the market (Z_v) increases. Fourth, a higher population (A_v) increases the land rent. Fifth, since the coefficients a_{3k} and a_{5k} are a function of β_k, population (A_v) and distance to the village (D_i) have a stronger impact on the land rent for crops with a higher labor elasticity. Similarly, the coefficient a_{4k} depends on α_k and δ_{1k} implying that the land rent for capital-intensive land-use systems or for crops with higher transport costs, is much more affected by changes in the accessibility to major markets (Z_v).

4. Empirical Estimation

Research Area

We estimate the above model for an area surrounding the Lore Lindu National Park in Central Sulawesi, Indonesia. The park is located in two districts comprising together five sub-districts and 119 villages. The rural households in these villages depend on agriculture for their livelihoods and cultivate a variety of crops including coffee, cocoa and coconut for the export market and rice for subsistence and for the local market. The national park is increasingly threatened by agricultural encroachment into the park's boundaries and unsustainable agricultural practices in the surrounding areas. This region is our research area and was chosen as a focus site for a collaborative research project.

Understanding the spatial land-use patterns and household behavior that determines these land-use patterns is extremely relevant for this forest-frontier area as the area is recognized as one the most important centers of endemic species in the world and forest protection is extremely important for conservation of the unique biodiversity (Waltert et al., 2003).

The Model

If a plot of land is allocated to the land use with the highest rent as expressed in equation (10), the probability of a plot i to be allocated to land use k can be expressed as:

$$\Pr(ik) = \frac{R_{ik}}{\sum_j R_{ij}} \tag{11}$$

Equation (10) and (11) describe a multinominal logit model that allows us to estimate the coefficients in equation (10) empirically. We estimate this multinominal logit model for the Lore Lindu region using a combination of socio-economic survey data and GIS data.

Data

To estimate the model we use a unique combination of socio-economic and geographic data. We use land-use data derived from satellite image interpretation, other geographic data derived from a variety of maps and socioeconomic data derived from a village-level survey and census data.

First, we use a land-use map derived from the interpretation of a Landsat ETM scene taken on August 24th, 2001 and covering the research area. The interpretation procedure used maximum likelihood classification techniques with ground-truth data, resulting in a land cover map with ten different land-use classes and a resolution of 15 by 15m. The identified land-use classes include open and closed forest, four classes of agricultural land use, grassland, reed, water, settlement areas, clouds and shadow.

Second, along with the land-use map, additional geographic information was compiled in a GID system. This includes a topographic map[1] that was digitized and used to construct a digital elevation model with a resolution of 70 by 70 m and an administrative map with sub-district boundaries, location of populated village centres, and demarcation of the National Park. In addition a

[1] This is derived from the 1991 edition of the Peta Rupabumi Indensia constructed by Bakosurtanal (Badan Koordinasi Survei dan Pemetaan Nasional – National Coordinating Agency for Surveys and Mapping). The scale is 1:50,000. The maps are based on aerial photographs of the years 1981,1982 and 1989.

detailed road map[2] distinguishing between asphalt roads, gravel or dirt roads and walking tracks is included. Also a soil map[3] providing rough information on different soil types was digitized and included in the GIS.

Third, 80 out of the 119 villages in the research area have been selected to be included in a comprehensive village-level survey based on a stratified random sampling method. The villages were classified in ten different strata based on the proximity to the National Park, the population density in the village and the share of migrants among the village population. We selected disproportionately more villages among the villages located close the National Park because the collaborative research project mainly concerns the stability of the rainforest margins. With respect to the other criteria, the selection was proportionate to population size.

The village survey was implemented in the period March-July 2001 using a formal quantitative questionnaire inquiring about the following topics: demographics, land use, agricultural production, agricultural technology, marketing, land and labor institutions, conservation issues, livestock holdings, infrastructure, and household well-being. The interviews were held in group discussions with village representatives and village elders. The village survey data were complemented with socio-economic data from secondary sources, including population and agricultural censuses for several years.

Combining Spatial and Non-Spatial Data

We estimate the model described by equation (10) and (11) using non-spatial socio-economic survey data and spatial geographic data. The way of combining spatial and non-spatial data requires some more explanation.

First, the resolution of our land-use map derived from the satellite image interpretation is 15×15 m^2 while the other grid-based data sources derived from the DEM have a resolution of 70×70 m^2. We take the smallest common multiple of these pixel sizes, 210×210 m^2 as the final resolution of the analysis. This resolution is used to construct new grids from vector data sources and to resample existing raster data. The grids correspond to a plot size of 4.4 hectares.

---

[2] The road map is constructed using information from the topographic map, a recent road map constructed by Bappeda (Badan Penencanaan Pembangunan Daereh – Directory of Development Planning Board) and own observations from the field.

[3] The 1:1,000,000 soil map was constructed by Bakosurtanal (Badan Koordinasi Survei dan Pemetaan Nasional – National Coordinating Agency for Surveys and Mapping) in 1995.

Second, because of a lack of data on village boundaries (the smallest administrative unit for which reliable maps exist is the sub-district) we need to construct artificial boundaries in order to be able to link village-level socioeconomic data and spaital data. We do so by constructing Thiessen polygons, polygons around each village center such that each location within a polygon is closer to the village center of that polygon than to any other village center. (see e.g. Müller and Zeller, 2002). The boundaries constructed in this way are certainly not to be interpreted as representing real administrative boundaries. Rather they are an artificial spatial unit for linking village-level data to spatial data.

Third, the survey covers only a sample of 80 of the 119 villages in the region. These 80 villages include a total of 104,085 pixels of 210×210 m². For the estimation of the model we take a non-random sub-sample of 55,464 observations, which consists of all pixels that are located less than three km from roads, walking tracks and village centers. The sub-sample includes 95 percent of the agricultural area and 47 percent of the forest area of the total sample. This represents agricultural and forest marginal areas where most land-use changes are taking place and which we are most interested in.

Fourth, some of the variables used in the model are spatially explicit and measured at the pixel level while others are measured at the village level. An overview of the variables used in the model is given in table 1. The dependent variable is derived from satellite image interpretation and constitutes a categorical variable with four different land-use types: forest, annual crops, perennial crops and grassland.

RESULTS

The results of the estimation of the multinominal logit model are reported in table 2. We estimate the model taking forest as the comparison land-use category and therefore the results should be interpreted as such. We indicate the estimated coefficients as well as the estimated relative risk ratio or odds ratio which are easier to interpret. The majority of the estimated effects is statistically significant and has the expected signs based on the refined von Thunen – Ricardo land rent model.

Table 1. Overview of the variables used in the multinominal logit model

Dependent Variable	Variable name	Categories		Freq.	Perc.	Scale
	LANDUSE	Forest		42,717	77.02%	Pixel
		Annual crops		4,165	7.51%	Pixel
		Perennial crops		4,225	7.62%	Pixel
		Grassland		4,357	7.86%	Pixel
Explanatory Variables	**Variable name**	**Mean**	**Std. Dev.**	**Minimum**	**Maximum**	**Scale**
Geophysical variables						
Slope (degrees)	SLOPE	10.79	8.03	0	45	pixel
Elevation (100 m)	ELEV	10.49	3.70	0.25	23.51	pixel
Aspect (degrees)	ASP	187	105	0	360	pixel
Aspect squared (degrees)	ASP^2	45,967	38,822	0	129,598	pixel
Distance to river (100 m)	TORIVER	1.15	1.11	0	9.60	pixel
Location variables						
Distance to road (100 m)	TOROAD	84.29	72.42	0	322.74	pixel
Distance to hamlet (km)	TOVILL	3.47	2.95	0	20.28	pixel
Dummy inside the National Park	PARK	0.33	0.47	0	1	pixel
Distance to city (km)	TOCITY	92	33	9	145	hamlet
Distance to city squared (km)	$TOCITY^2$	9,527	5,937	77	21,091	hamlet
Distance to district capital (km)	TODISTRICT	27	19	0	67	hamlet
Socioeconomic variables						
# years irrigation infrastructure	IRR_YRS	3.80	9.53	0	71	village
ln (population)	lnPOP	6.64	0.67	5.46	8.45	village
% of population in working age	POPWORK	67	8	39	83	village
Spatially lagged variables						
Lagged slope (degrees)	LAG	10.76	6.85	0	35	pixel
Lagged slope * slope	SLOPE_LAG	211.49	222.92	0	1,680	pixel

Source: Own calculations.

The probability to find annual crops relative to forest is higher on flatter plots with a higher elevation, on less northward sloping hillsides, on plots closer to rivers and village centers, and outside the National Park. The same is true for perennial crops but the effects of slope and distance to rivers are a lot smaller and not statistically significant. The probability of a plot being allocated to agricultural land use, annual or perennial crops, is higher in villages closer to cities and district capitals and in villages with a larger population and work force.

In addition, access to irrigation decreases the likelihood of perennial crops relative to forest cover. The probability to find grassland relative to forest is higher on plots at a lower altitude, further away from roads and in villages closer to cities and district capitals.

Table 2. Results of the multinominal logit model (with forest as comparison)

Number of observations	55,464
Wald chi2(42)	12,945
Prob > chi2	0.000
Pseudo R^2	0.380
Log Likelihood	-27,211

	Relative Risk Ratio	Coeffcient	Robust Std. Err.	z	P>|z|
Annual Crops					
LAG	0.8788	-0.1292	0.0123	-10.52	0.000
SLOPE	0.9410	-0.0608	0.0244	-2.50	0.013
SLOPE_LAG	1.0023	0.0023	0.0013	1.74	0.082
ELEV	0.8047	-0.2173	0.0357	-6.09	0.000
ASP	0.9939	-0.0061	0.0020	-3.14	0.002
ASP²	1.0000	0.0000	0.0000	2.11	0.035
TORIVER	0.7828	-0.2449	0.0910	-2.69	0.007
TOROAD	0.9982	-0.0018	0.0018	-1.03	0.304
TOVILL	0.5626	-0.5752	0.1183	-4.86	0.000
TOCITY	0.9351	-0.0671	0.0113	-5.93	0.000
TOCITY²	1.0005	0.0005	0.0001	5.82	0.000
TODISTRICT	0.9862	-0.0139	0.0087	-1.61	0.108
PARK	0.2348	-1.4489	0.2325	-6.23	0.000
lnPOP*	1.2459	0.2199	0.1688	1.30	0.193
POPWORK	1.0206	0.0204	0.0106	1.92	0.054
IRR_YRS	1.0048	0.0047	0.0070	0.68	0.497
Constant		2.9431	1.3890	2.12	0.034
Perennial Crops					
LAG	0.9565	-0.0445	0.0121	-3.68	0.000
SLOPE	0.9760	-0.0243	0.0175	-1.39	0.165
SLOPE_LAG	1.0014	0.0014	0.0007	2.02	0.043
ELEV	0.7016	-0.3544	0.0304	-11.65	0.000
ASP	0.9864	-0.0137	0.0018	-7.44	0.000
ASP²	1.0000	0.0000	0.0000	4.88	0.000
TORIVER	0.9931	-0.0070	0.0907	-0.08	0.939
TOROAD	1.0008	0.0008	0.0016	0.52	0.606
TOVILL	0.6176	-0.4819	0.0754	-6.39	0.000
TOCITY	0.9383	-0.0637	0.0098	-6.48	0.000
TOCITY²	1.0004	0.0004	0.0001	5.55	0.000
TODISTRICT	0.9921	-0.0079	0.0080	-0.98	0.325
PARK	0.4452	-0.8093	0.1892	-4.28	0.000
lnPOP*	1.3876	0.3276	0.1835	1.79	0.074
POPWORK	1.0095	0.0095	0.0111	0.85	0.393
IRR_YRS	0.9827	-0.0175	0.0060	-2.93	0.003
Constant		3.7524	1.3486	2.78	0.005
Grassland					
LAG	0.8121	-0.2082	0.0215	-9.67	0.000
SLOPE	0.9841	-0.0160	0.0286	-0.56	0.576
SLOPE_LAG	1.0018	0.0018	0.0016	1.10	0.270
ELEV	0.9086	-0.0958	0.0452	-2.12	0.034
ASP	0.9932	-0.0068	0.0019	-3.59	0.000
ASP²	1.0000	0.0000	0.0000	3.70	0.000
TORIVER	1.0055	0.0055	0.0692	0.08	0.937
TOROAD	0.9941	-0.0059	0.0025	-2.37	0.018
TOVILL	1.0473	0.0463	0.0497	0.93	0.352
TOCITY	0.9809	-0.0193	0.0218	-0.88	0.377
TOCITY²	1.0003	0.0003	0.0002	2.02	0.043
TODISTRICT	0.9767	-0.0236	0.0161	-1.47	0.142
PARK	0.1164	-2.1505	0.3463	-6.21	0.000
lnPOP*	0.8251	-0.1923	0.3801	-0.51	0.613
POPWORK	1.0208	0.0206	0.0215	0.96	0.337
IRR_YRS	0.9974	-0.0026	0.0162	-0.16	0.874
Constant		0.6011	3.1487	0.19	0.849

* Instrumented variable

Source: own estimations.

Implications of the Empirical Model

The results of the multinominal logit model indicate that geophysical land characteristics are very important factors in determining the spatial land-use pattern. Other spatially explicit land-use studies also found highly significant effects of topographic features and other geophysical land characteristics on the probability of certain land-use types or land-use changes (e.g. Nelson and Hellerstein, 1997; Cropper *et al.*, 2001; Deininger and Minten, 2002; Müller and Zeller, 2002). The estimated effect of slope and lagged slope on the probability of agricultural land use relative to forest is negative while the interaction term of slope and lagged slope has a positive effect. This means that agriculture is found more often on flatter plots and in less mountainous surroundings. Yet, the probability to find agricultural land on steeper plots is higher in more mountainous surroundings. Further, annual crops are found closer to rivers and on less steep slopes, which indicates that (potential) agricultural yields of annual crops are determined to a large extent by topographic characteristics and access to water. The likelihood of perennial crops relative to forest is less influenced by the slope of the plot. In addition, the estimated effects show that agricultural land use becomes less likely and forest more likely with increasing elevation. The results demonstrate that differences in Ricardian land rent are very important in explaining the present land-use pattern.

The location of a plot in relation to villages and roads is hypothesized to be crucial in determining land use. The results show that the distance to populated centers has a much larger effect on land use than the distance to roads, which is statistically not significant[4]. For each additional five kilometer away from a village center, it is 18 times less likely to find annual crops and 11 times less likely to find perennial crops compared to forest[5]. Further, the accessibility to cities and major towns might matter. We find that with every ten kilometer distance between the village center and the district capital, it becomes 1.15 times less likely to find annual crops and 1.08 times less likely

---

[4] To exclude the possibility that the effect of distance to roads is statistically not significant because of multicollinearity problems, we estimated the multinominal regression dropping all other cost-of-access variables that are correlated to some extent with distance to roads. The estimated effects of distance to roads and the level of significance did not change much, which demonstrates that there is no multicollinearity problem for the variable.

[5] The variable *TOVILL* is measured in km and the estimated coefficients for this variable are –0.5752 and –0.4819 for annual crops and perennial crops respectively. So, the odds ratio for a five units (= 5 km) change is 1/exp (-0.5752*5)=18 for annual crops and 1/exp (-0.4819*5)=11 for perennial crops. Other odds ratios can be calculated in similar way.

to find perennial crops compared to forest. It is almost two times less likely to find perennial or annual crops on plots in villages located ten kilometer further away from the city. These results indicate that the distance from plots to village centers is much more important in determining land use than the distance from villages to towns and markets.

As in other spatially explicit economic land-use studies, the model for the Lore Lindu region shows that the attributes and the location of plots influence land use. However, most studies find a much larger impact of access to roads (e.g. Deininger and Minten, 2002; Müller and Zeller, 2002). Our results indicate that the location of plots with respect to roads is not important at all in determining the land-use type. Rather, the location of plots in relation to village centers and to a lesser extent the access of villages to markets matters. The land-use pattern in the Lore Lindu region is centered on villages rather than around roads and major markets, which relates to the history of the area. Villages might have a long history of establishment while roads were built more recently to connect villages. The spatial land-use pattern is quite different than for instance in the Amazon regions, which are characterized by road colonization implying clearance and settlement of forested areas after roads have been built.

The land-use pattern is centered on villages with annual crops cultivated closer to the village and perennial crops further away at forest margins. This could be related to the fact that village centers are located in flatter areas, which are more suitable for the cultivation of annual crops. However, differences in topographic features are accounted for in the model. Also in villages without much topographic variation the same land-use pattern emerges. The relation between distance to the village and the likelihood to find annual respectively perennial crops is likely associated with differences in labor intensity. Land-use systems based on the cultivation of annual crops are more labor intensive than systems with perennial crops. More labor-intensive land-use systems are located closer to village centers and less labor-intensive systems further away because this reduces the time spent to reach the fields.

The variable expressing distance to the city and its square have an opposite sign with an inflection point around 70 kilometer. This implies a U-shaped relation between distance to the city and the likelihood of agriculture land use relative to forest. The interpretation of this result is not straightforward and should be done with some caution. On the one hand, in villages closer than 70 kilometer from the city, the probability of agriculture decreases with distance to the city at an increasing rate. Since villages closer to the cities are better connected to the road network this could mean that the cost-of-access rather than the Euclidian distance between villages and cities matters. Better access to markets and lower transport costs increase the profitability of farming and increase the likelihood of agriculture relative to forest. On the other hand, in villages further than 70 kilometer from the city, the probability of agriculture increases with distance to the city. This could imply that in villages further from the market agriculture is more extensive, resulting in a lower probability of forest cover. In addition, remote villages are also more mountainous such that agriculture and forest compete for less suitable land. Further, we find that the probability to find annual crops relative to forest is much more influenced by the distance to district capital towns than is the case for perennial crops. Concerning the accessibility to major cities, similar effects are found for both land-use types. Annual crops constitute food crops such as rice and corn while perennial crops are usually export crops such as coffee and cocoa. Hence, the results suggest that local markets in the district capital towns are more important for trade of food crops while cities are important for the marketing of both food crops and export crops. The original von Thunen model and its application in spatial land-use models emphasize the importance of differences in transport costs for different crops in determining the spatial pattern of land use. Since the effect of distance to the city is not different for annual crops than for perennial crops, we do not find much evidence in support of this idea. The differences in transport costs between rice, the major annual crop, and cocoa beans, the major perennial crop product, might not be that large because these are both bulk products, which are quite easily transported.

Further, the results demonstrate that inside the Lore Lindu National Park it is 4.3 times less likely to find annual crops, 2.2 times less likely to find perennial crops and 8.6 times less likely to find grassland compared to forest. The forest inside the National Park is less likely to be cleared, which suggests that the establishment of a National Park is to some extent effective for forest conservation. Yet, the forest inside the National Park is more likely to be cleared for the cultivation of perennial crops than for the cultivation of annual crops. There are several possible explanations for this observation. First, the borders of the National Park are set along topographic features and the land inside the Park is less suitable for the cultivation of annual crops. Second, the risk of being caught (and fined) is less because plots with perennial crops inside the forest are less conspicuous. Third, the risk of being caught while working on the plot is less because perennial crops are less labor demanding. Fourth, it is easier to avoid fines by claiming to have planted perennial crops already before the demarcation of the National Park.

Next, we find that access to an improved irrigation system reduces the probability of perennial crops compared to forest. This implies that technical progress for paddy rice cultivation reduces pressure on forests. The results show that the likelihood of agricultural land use compared to forest increases with population and the share of workers in the population. A one percent increase in village population increases the probability of perennial crops and annual crops relative to forest with 33 and 22 percent respectively. With every five percent increase in the share of workers among the village population, it becomes 1.1 times more likely to find annual crops and 1.04 times more likely to find perennial crops relative to forest. Excluding the effects that are statistically not significant, these results point to the importance of population in shaping the spatial land-use pattern. There is not much evidence for a substantial effect of population and other socioeconomic variables in spatial land-use models in the literature. However, the effect might have been obscured due to the difference in aggregation level between land-use data at the pixel level and socioeconomic data at a much more aggregated level. For example, Cropper *et al.* (2001) and Deininger and Minten (2002) included district-level population into a spatial land-use model and find no significant effect. The aggregation bias might have led to an erroneous conclusion about the effect of population and other socioeconomic factors on land-use change and deforestation. Our model using village-level data might be better suited to elucidate the effects of socioeconomic factors on spatial land-use patterns.

CONCLUSION

We can conclude that the refinement of the von Thunen – Ricardo model to let prices (especially relative wages) be determined not solely by transport costs to a major output market, has provided a good basis for a description of the spatial land-use pattern in the Lore Lindu region. Land use is very much determined by differences in geophysical land characteristics and is centered around villages with labor-intensive land-use systems closer to populated centers. Differences in population levels, the available technology and the location of major markets and towns further shape the spatial land-use pattern.

REFERENCES

Bakosurtanal (Badan Koordinasi Survei dan Pemetaan Nasional). 1991. Peta Rupabumi Indonesia skala 1:50,000. Jakarta, Bakosutanal.

Bakosurtanal (Badan Koordinasi Survei dan Pemetaan Nasional). 1995. Peta Dasar. Digetasi dari Petah Rupabumi. Skala 1:1,000,000. Jakarta, Bakosutanal.

Bappeda (Badan Perencanaan Pembangunan Daerah Kabupaten Donggala dan Kantor Kabupaten Donggala). 2000. Analisa Peta Dasar Neraca Lahan Kabupaten Donggala. Palu.

Chomitz, K.M. and D. Gray. 1996. Roads, Lands Use, and Deforestation: A Spatial Model Applied to Belize. *World Bank Economic Review*. 10: 487-512.

Cropper, M., J. Puri, and C. Griffiths. 2001. Predicting the Location of Deforestation: The Role of Roads and Protected Areas in North Thailand. *Land Economics*. 77: 172-186.

Deininger, K. and B. Minten. 1999. Poverty, Policies, and Deforestation: The Case of Mexico. *Economic Development and Cultural Change*. 47:313-343.

Fox, J., R. Kanter, S. Yarnasarn, M. Ekasingh, and R. Jones. 1994. Farmer Decision Making and Spatial Variables in Northern Thailand. *Environmental Management*. 18: 391-399.

Gobin, A., P. Campling, and J. Feyen. 2002. Logistic Modelling to Derive Agricultural Land Use Determinants: A Case Study from Southeastern Nigeria . *Agriculture, Ecosystems and Environment*. 89: 213-228.

Irwin, E.G. and J. Geoghegan. 2001. Theory, Data, Methods: Developing Spatially Explicit Economic Models of Land Use Change. *Agriculture, Ecosystems and Environment*. 85: 7-23.

Lambin, E.F., M.D.A. Rounsevell, and H.J. Geist. 2000. Are Agricultural Land-Use Models Able to Predict Changes in Land-Use Intensity? *Agriculture, Ecosystems and Environment*. 82:321-331.

Liu, D.S., L.R. Iverson, and S. Brown. 1993. Rates and Patterns of Deforestation in the Philippines: Application of Geographic Information System Analysis. *Forest Ecology and Management*. 57:1-16.

Mertens, B. and E.F. Lambin. 1997. Spatial Modelling of Deforestation in Southern Cameroon - Spatial Desegregation of Diverse Deforestation Processes. *Applied Geography*. 17:143-162.

Mertens, B. and E.F. Lambin. 2000. Land-Cover-Change Trajectories in Southern Cameroon. *Annals of the Association of American Geographers*. 90:467-494.

Munroe, D., J. Southworth, and C.M. Tucker. 2002. The Dynamics of Land-Cover Change in Western Honduras: Exploring Spatial and Temporal Complexity. *Agricultural Economics*. 27:355-369.

Müller, D. and M. Zeller. 2002. Land Use Dynamics in the Central Highlands of Vietnam: A spatial Model Combining Village Survey Data and Satellite Imagery Interpretation. *Agricultural Economics*. 27:333-354.

Nelson, G.C. 2002. Introduction to the Special Issue on Spatial Analysis for Agricultural Economists. *Agricultural Economics*. 27:197-200.

Nelson, G.C. and J. Geoghegan. 2002. Deforestation and Land Use Change: Sparse Data Environments. *Agricultural Economics*. 27:201-216.

Nelson, G.C., V. Harris, and S.W. Stone. 2001. Deforestation, Land Use, and Property Rights: Empirical Evidence from Darien, Panama. *Land Economics*. 77:187-205.

Nelson, G.C. and D. Hellerstein. 1997. Do Roads Cause Deforestation? Using Satellite Images in Econometric Analysis of Land Use. *American Journal of Agricultural Economics*. 79:80-88.

Serneels, S. and E.F. Lambin. 2001. Proximate Causes of Land-Use Change in Narok District, Kenya: A Spatial Statistical Model. *Agriculture, Ecosystems and Environment*. 85:65-81.

Vance, C. and J. Geoghegan. 2002. Temporal and Spatial Modelling of Tropcial Deforestation: a Survival Analysis Linking Satellite and Household Survey Data. *Agricultural Economics*. 27:317-332.

Verburg, P.H., Y. Chen, and T. Veldkamp. 2000. Spatial Explorations of Land Use Change and Grain Production in China. *Agriculture, Ecosystems and Environment*. 82:333-354.

Verburg, P.H., T. Veldkamp, and J. Bouma. 1999. Land Use Change under Conditions of High Population Pressure: The Case of Java. *Global Environmental Change*. 9:303-312.

Verburg, P.H. and A. Veldkamp. 2001. The Role of Spatially Explicit Models in Land-Use Change Research: A Case Study for Cropping Patterns in China." *Agriculture, Ecosystems and Environment*. 85:177-190.

von Thünen, J.H. 1826. *Der Isolierte Staat in Beziehung auf Landwirtschaft und Nationaloekonomie*.

Waltert, M., M. Langkau, M. Maertens, M. Härtel, S. Erasmi, and M. Mühlenberg. 2003. Predicting the Loss of Bird Species from Deforestation in Central Sulawesi. In G. Gerold, M. Fremery, and E. Guhardja, editors, *Land Use, Nature Conservation and the Stability of Rainforest Margins in Southeast Asia*. Springer. Berlin.

In: New Developments in Biodiversity … ISBN: 978-1-61324-374-9
Editor: Thomas W. Pace

Chapter 2

UTILIZATION OF BOVIDS IN TRADITIONAL FOLK MEDICINE AND THEIR IMPLICATIONS FOR CONSERVATION

Rômulo Romeu da Nóbrega Alves*[*,1], Raynner Rilke Duarte Barboza[2], Wedson de Medeiros Silva Souto[3] and José da Silva Mourão[1]

[1]Departamento de Biologia, Universidade Estadual da Paraíba, Avenida das Baraúnas, Campina Grande, Paraíba 58109-753, Brasil
[2]Pós-Graduação em Ciência e Tecnologia Ambiental, Universidade Estadual da Paraíba, Avenida das Baraúnas, Campina Grande, Paraíba 58109-753, Brasil
[3]Programa Regional de Pós-Graduação em Desenvolvimento e Meio Ambiente (PRODEMA), Universidade Estadual da Paraíba, Avenida das Baraúnas, Campina Grande, Paraíba 58109-753, Brasil

ABSTRACT

Animals and products derived from different organs of their bodies have constituted part of the inventory of medicinal substances used in various cultures since ancient times. Regrettably, wild populations of

* E-mail: romulo_nobrega@yahoo.com.br.

numerous species are overexploited around the globe, the demand created by the traditional medicine being one of the causes of the overexploitation. Mammals are among the animal species most frequently used in traditional folk medicine and many species of bovids are used as medicines in the world. The present work provides an overview of the global usage of bovids in traditional folk medicine around the world and their implications for conservation. The results demonstrate that at least 55 bovids are used in traditional folk medicine around the world. Most of species (n=49) recorded were harvested directly from the wild, and only six species of domestic animals. Of the bovids recorded, 50 are included on the IUCN Red List of Threatened Species and 54 are listed in the CITES. By highlighting the role played by animal-based remedies in the traditional medicines, we hope to increase awareness about zootherapeutic practices, particularly in the context of wildlife conservation.

Keywords: bovids, traditional medicine, wildlife conservation

INTRODUCTION

Traditional medicinal systems play a key role in health care around the world. The World Health Organization has estimated that 60–80% of the population of non-industrialized countries rely on traditional healthcare for their basic health care needs, either on its own or in conjunction with modern medical care. Much of the world's population depends on traditional medicine, especially within developing countries and the demand for traditional medicine is increasing in many countries [1].

A tremendous variety of plants and animals are used in the preparation of traditional medicines. The pharmacopoeia of folk societies as well as of traditional (such as those of the Chinese, Ayurvedic, Unani) and western medical systems contain thousands of uses for medicines made from leaves, herbs, roots, bark, animals, mineral substances and other materials found in nature [1-4].

Since immemorial times animals and products derived from different organs of their bodies have constituted part of the inventory of medicinal substances used in many cultures, and such uses still exist in ethnic folk medicine [4]. Testimony to the medical use of animals began to appear with the invention of writing. Archives, papyruses, and other early written historical sources dealing with medicine, show that animals, their parts, and their products were used for medicine [5].

Bovids form one of the most prominent families of herbivores and constitute an important group for human beings, most of all when it comes to food, economy and religion. In traditional folk medicine, Bovids are among the mammals species most frequently used in different localities along the world. In recent years, several works reported the use of this group species for medicinal purposes [6-13] justifying the cultural lore importance of that group in different human societies, especially in traditional folk medicine.

From the historical point of view, a variety of interactions established between humans and bovids contributed, in higher or less scale, for the development of several cultures as well as the human species survival. If in a prime moment the hunt of several animals such antelopes, gazelles, buffalos and many others bovid species settled one of the prominent manners of food obtaining, equally important was the animals' domestication to the sedentarization and formation of the first urban nucleus occurred during the Holocene (Neolithic) [14-16]. Goats and sheep are among the earliest animals domesticated in southern and southwestern Asia during the early Holocene [15, 17, 18] and many evidences indicate that the first urban nucleus appeared in Asia are directly related to pastoral activities of flocks breeding (see Ponting [14]).

Unfortunately, many bovids species are found in Lists of Threatened Species, and for some of them, one of the reasons is the use and exploitation by humans. Currently, the family Bovidae includes 143 species in 50 genera represented by goats, sheep, gazelles, antelopes, goat-antelopes and cattle [19]. Because of such hunting, combined with loss of suitable habitat and other causes, most species are now considered to be endangered in the wild. According with IUCN [20], a large number of mammals are used in traditional folk medicine and many of these are of relevant to conservation. Understanding that the use of animals for medicinal purposes is part of a body of traditional knowledge which is increasingly becoming more relevant to discussions on conservation biology, public health policies, sustainable management of natural resources, biological prospection, and patents [4]. In this context, this study has focused on the global use of bovids in traditional folk medicine around the world and the implications for its conservation. We hope this chapter will serve as stimulus for further research about this use of biodiversity and its implications for bovids's conservation.

METHODS

In order to examine the diversity of bovids used in traditional medicine, all available references or reports of folk remedies based on bovids sources were examined. Only taxa that could be identified to species level were included in the database. Scientific names provided in publications were updated according to the ITIS Catalogue of Life: 2008 Annual Checklist [21] and Mammal Species of the World (MSW) [19]. The conservation status of the bovids species follows IUCN [20] and CITES [22]. All the bovids sources used in traditional medicine were placed in references.

A total of 55 studies, official documents and other types of reliable publications supplied the information analyzed: Lev [5, 23], Mahawar and Jaroli [7, 8], Alves et al. [9], El-Kamali [10], Alves [11], Kakati et al. [12], Negi and Palyal [13], Adeola [24], Alakbarli [25], Almeida and Albuquerque [26], Alves and Rosa [27-29], Alves et al. [30], Apaza et al. [31], Australian Department of Environment and Heritage [32], Barboza et al. [33], Bensky et al. [34], Bryan [35], Cameron [36], Caprinae Specialist Group [37], Carpaneto and Germini [38], Chan [39], CITES [40-42], Costa-Neto [43-45], Costa-Neto and Oliveira [46], Dedeke et al. [47], Duckworth et al. [48], Enqin [49], Estes [50], Fischer and Linsenmair [51], Fitzgerald [52], Green [53], Green [54], Guojun and Luoshan [55], Homes [56, 57], Lavigne et al. [58], Lev and Amar [59], Li [60], Malik et al. [61], Marques [62], Marshall [63], Monroy-Vilchis et al. [64], Nilsson [65], Nunn [66], Pieroni et al. [67], Powell[68], Pujol [69], Rajchal [70], Ritter [71], Schaller [72], Schroering [73], Sheng [74], Silva et al. [75], Simelane and Kerley [76], Sodeinde and Soewu [77], Soewu [78], Solanki and Chutia [79], Souza [80], Stetter [81], The Chinese Materia Medica Dictionary [82], Thompson [83], Tierra [84], True [85], Truong et al. [86], *Vázquez* et al. [87], Walston [88], Zhang [89].

RESULTS AND DISCUSSION

Bovids were hunted by humans from very early times and have been used for different purposes. The frequent depiction of bovids species in rock drawings attests to their importance to early hunters. Pre-historic societies used bovids and their products (primarily consumed as food), and the use of this animals has perpetuated throughout the history of humanity. In contemporary societies, bovids wild animals are used for a wide variety of finalities, such as food resources, pets, cultural activities, for medicinal and magic-religious purposes, and their body parts or sub-products are used or sold as clothing and tools. Bovids have been depicted in art and have played important roles in mythology and religion. The American bison is seen to have great spiritual importance by many Native Americans. Cattle have played an important part in many cultural and religious traditions, as they continue to do in Hinduism today. In Judaism, Islam, and Christianity, sheep and goats have symbolic roles. In Judaism, a shofar, a "trumpet" made from the horn of a ram or sometimes that of a goat or antelope, is sounded at Rosh Hashanah, the celebration of the New Year [90]. In Christianity, it is believed that the sounding of a shofar will announce the return of Christ.

The roles that bovids play in folk practices related to the healing and/or prevention of illnesses have been recorded in different social-cultural contexts worldwide. As evidenced in the present review, at least 56 species (39% of described bovids species) belonging to 29 genera and 8 subfamilies are used in traditional folk medicine. The subfamily with the largest numbers of species used were Caprinae (with 20 species), followed by Bovinae (14), and Antilopinae (9) (Table 1). These results were expected once the mentioned subfamilies are the most numerous in terms of bovid species (see Wilson and Reeder [19]).

Despite the fact that studies recording the use of bovids in traditional medicine are all relatively recent, an analysis of historical documents indicated that bovids have been used in traditional medicines since ancient times. Historical sources of ancient Egypt and of several civilizations of ancient Mesopotamia, mainly the Assyrian and the Babylonian, mention the medicinal uses of substances derived from bovids, for example, cattle milk, goat's skin, gazelle sinew, even sheep, and the glands of the musk deer [35, 50, 66, 68, 71, 81, 83]. Musk is used as a medicinal in Islamic countries, India and countries of the Far East [53]. Its use in medicine was recorded by Aetius, the Greek physician, in circa 520 AD (Pereira, 1857 quoted by Green [53]), and in China and India musk has been considered a superior medicine since the fifth century AD [53]. However, as early as the Han Dynasty (200 BC – 200 AD), records in Shennong Bencao Jing show the use of musk in traditional Chinese medicine (TCM) [70].

Some widespread species are used in different countries like *Naemorhedus sumatraensis* (Bechstein, 1799) in China and Lao PDR [48, 82] and probably in most of Asian Southeast countries; *Bubalus bubalis* (Linnaeus, 1758) and *Saiga tatarica* (Linnaeus, 1766) in China and India [7, 34, 84], *Ovis aries* (Linnaeus, 1758) in Brazil and Sudão [10, 27-29, 33], *Bos taurus* Linnaeus, 1758 and *Bos frontalis* Lambert, 1804 in Brazil and India [7, 8, 27-30, 44, 45].

Table 1. Felidae used in the wordwide traditional medicine

BOVIDAE	IUCN 2008 Red List	2008 CITES Appendices
Subfamily Aepycerotinae		
Aepyceros melampus (Lichtenstein, 1812)	LC	
Subfamily Alcelaphinae		
Alcelaphus buselaphus (Pallas, 1766)	LC	
Connochaetes gnou (Zimmermann, 1780)	LC	
Subfamily Antilopinae		
Antilope cervicapra (Linnaeus, 1758)	NT	III (Nepal)
Gazella dorcas (Linnaeus, 1758)	VU	III (Algeria, Tunisia)
Gazella subgutturosa (Güldenstaedt, 1780)	VU	
Neotragus moschatus (Von Dueben, 1846)	LC	
Oreotragus oreotragus (Zimmermann, 1783)	LC	
Procapra gutturosa (Pallas, 1777)	LC	
Procapra picticaudata Hodgson, 1846	NT	
Raphicerus melanotis (Thunberg, 1811)	LC	
Saiga tatarica (Linnaeus, 1766)	CR	II
Subfamily Bovinae		
Bison bison (Linnaeus, 1758)	NT	II
Bos frontalis Lambert, 1804		
Bos grunniens Linnaeus, 1766		
Bos javanicus d'Alton, 1823	EN	
Bos sauveli Urbain, 1937	CR	I
Bos taurus Linnaeus, 1758		
Bubalus bubalis (Linnaeus, 1758)		
Bubalus depressicornis (H. Smith, 1827)	EN	I
Pseudoryx nghetinhensis Dung et al., 1993	CR	I
Syncerus caffer (Sparrman, 1779)	LC	
Taurotragus oryx (Pallas, 1766)		
Tragelaphus euryceros (Ogilby, 1837)		
Tragelaphus scriptus (Pallas, 1766)	LC	
Tragelaphus strepsiceros (Pallas, 1766)	LC	
Subfamily Caprinae		
Ammotragus lervia (Pallas, 1777)	VU	II
Budorcas taxicolor Hodgson, 1850	VU	II
Capra falconeri (Wagner, 1839)	EN	I

BOVIDAE	**IUCN 2008 Red List**	**2008 CITES Appendices**
Capra hircus Linnaeus, 1758		
Capra ibex Linnaeus, 1758	LC	
Capra sibirica (Pallas, 1776)	LC	
Hemitragus hylocrius (Ogilby, 1838)	EN	
Hemitragus jemlahicus (H. Smith, 1826)	NT	
Naemorhedus baileyi Pocock, 1914	VU	I
Naemorhedus caudatus (Milne-Edwards, 1867)	VU	I
Naemorhedus crispus (Temminck, 1845)		
Naemorhedus goral (Hardwicke, 1825)	NT	I
Naemorhedus sumatraensis (Bechstein, 1799)	VU	I
Naemorhedus swinhoei (Gray, 1862)		
Ovis ammon (Linnaeus, 1758)	NT	I, II
Ovis aries (Linnaeus, 1758)		
Ovis canadensis Shaw, 1804	LC	II (Only the population of Mexico)
Ovis vignei Blyth, 1841		I, II
Pantholops hodgsonii (Abel, 1826)	EN	I
Pseudois nayaur (Hodgson, 1833)	LC	
Subfamily Cephalophinae		
Cephalophus jentinki Thomas, 1892	EN	I
Cephalophus maxwellii (H. Smith, 1827)	LC	
Cephalophus monticola (Thunberg, 1789)	LC	II
Cephalophus natalensis A. Smith, 1834	LC	
Cephalophus rufilatus Gray, 1846	LC	
Sylvicapra grimmia (Linnaeus, 1758)	LC	
Subfamily Hippotraginae		
Hippotragus equinus (Desmarest, 1804)	LC	
Subfamily Reduncinae		
Kobus ellipsiprymnus (Ogilby, 1833)	LC	
Kobus kob (Erxleben, 1777)	LC	
Kobus leche Gray, 1850	LC	II

Legend: IUCN Red List of Threatened Species Categories – CR (Critically Endangered), DD (Data Deficient), EN (Endangered), EX (Extinct), EW (Extinct in the Wild), LR/cd (Lower risk: Conservation dependent), LR/lc or LC (Least Concern), LR/nt or NT (Near Threatened), VU (Vulnerable). CITES Appendices – I, II or III.

A same species can be used in the treatment of different diseases. For instance, the South-east musk continues to be a popular Traditional Chinese Medicine for the effective treatment of improving blood circulation and relieving ailments of the heart, nerves and breathing system. It can revive unconscious patients, stimulate circulation of vital energy and blood and it also possesses anti-inflammatory and analgesic effects. It is used in the treatment of delirium, stroke, unconsciousness, miscarriage, ejection of stillborn fetus, acute angina pectoris, acute abnormal pain, skin infection, sore throat, sprained joints, trauma, and paralysis. It occurs in three dominant forms, as oils or sprays, medicated plasters and raw musk powder [91]. The musk powder can be dissolved into water then used as a salve on parts of the body that suffer rheumatic pains. Musk powder can also be dissolved into wine or ginseng tea and consumed to improve the body's blood circulation. Preparation of musk can be used for both external and internal applications. Musk oils can be brought 'off the shelf' or blended with snake bile or various plant herbs. These are used to obtain immediate and short-term relief of rheumatic and muscular pains caused by blood stasis [70].

The domestic cattle - *Bos taurus*, perhaps the most well-known of the bovids, is an important source of remedies in traditional medicine. In Brazil, for instance, bone marrow, horns, medulla and penis of *B. taurus* are indicated for treatment of alcoholism, rib cage pain, male impotence, anaemia, dizziness, flu, pneumony, rheumatism, thrombosis, cough, asthma, sinusitis, sore throat, wounds and for removal of thorns [27-30, 44-46, 75]. In India, weakness due to fever is cure by drinking domestic cattle's urine and Ghee with black pepper are given orally to neutralize snake poison [8]. Ox gall was listed among the animals' drugs originated in Europe (see True [85]). In Viet Nam, gall, gall-stone and horn of *Bos taurus* are used in local folk medicine [92].

The effectiveness of most of the medicines from wild animals and their by-products has not been scientifically studied and proven and their potency in many cases may be questionable. As pointed out by Pieroni et al. [93], the chemical constituents and pharmacological actions of some animal products are already known to some extent and ethnopharmacological studies focused on animal remedies could be very important in order to clarify the eventual therapeutic usefulness of this class of biological remedies. However, research with therapeutic purposes into the products of the animal kingdom has been neglected until recently [4]. Another aspect that needs to be emphasized is that sanitary conditions of the zootherapeutics products generally were poor with obvious contamination risks to these products [9, 27, 29]. These observations point to the need for sanitary measurements to be taken with medicinal animal products and the importance of including considerations about zootherapy into public health programs.

Most of the species used (n = 51; 91%) are wild. In most cases remedies were prepared from dead specimens. Many of the medicinal animals are of conservation concern. Most of the recorded species (46 of 56) are included on the IUCN Red List of Threatened Species [20] (with 7 species classified as vulnerable and 6 are endangered, besides 1 critically endangered) and 21 are listed in the CITES list (Convention on International Trade in Endangered Species of Wild Fauna and Flora [22] (see Table 1).

Ingredients derived from these biological resources are not only widely used in traditional remedies, but are also increasingly valued as raw materials in the preparation of modern medicines and herbal preparations. As a result, harvest and trade of species for traditional medicines may pose a threat to their survival [94]. Traditional Chinese Medicine, for example, relies on animal and plant substances as raw ingredients for prescription medicine and manufactured patent pharmaceuticals [91].

Musk is a good example. It is considered one of the most frequently used animal products in traditional medicine practices [36]. The musk secreted by the musk gland of the males has been used in the perfumery industries for a long time for its intensity, persistence and fixative properties. In Asia, including China, it has also long been used in traditional medicine as a sedative and as a stimulant to treat a variety of ailments [53, 56, 74]. In China, Musk deer have been hunted for musk, and musk purchasing has been conducted in rural markets or via local medicine companies and the perfume industry perfume is produced based on natural musk, but production is not high at present [89]. The raw ingredients may be used directly after some preparation (grinding, washing, boiling, drying etc.) or may be made into factory processed forms such as plasters, pills or tablets and packaged in mass quantities for national or worldwide distribution.

In China, the effects of musk (*Moschus chrysogaster*) have been known in Traditional Chinese Medicine (TCM) for several thousand years, musk being used in about 300 pharmaceutical preparations [95]. China has a high domestic demand for musk [56], and this originates from both legal and illegal sources within the country. The total demand for musk is between 500 to 1,000 kg per year in China [74].

Musk (*Moschus* spp.) is currently used in as many as 400 Chinese and Korean traditional medicines to treat ailments of the circulatory, respiratory and nervous systems [96]. The demand for musk from China alone is between

500-1000 kg per annum needing extraction from pods of more than 100,000 male musk deer [96]. In Russia, 400-500 kg of raw musk was traded illegally between 1999 and 2000, needing extraction from 17,000-20,000 stags [57]. In 1987, 800 pounds of musk worth $14 million were smuggled out of China, the product of 53,000 male deer. More than 100,000 deer had been killed in the quest for these glands, since many of the dead deer were females and young which were discarded. The glands were exported to Japan [72]. An average of 700 pounds of musk are sold in world markets each year, much of it going to Hong Kong, the international center for musk; Japan is a major consumer, using musk to treat a variety of illnesses [52]. Between 1974 and 1983, Japan imported between 250 and 700 pounds of musk per year, worth an average of $4.2 million; imports increased in 1987 to 1,800 pounds, an all time high, and sold for $32,468 a pound [52].

The horns of Saiga antelope (*Saiga tatarica*), a species being slaughtered by the hundreds of thousands in Russia and Central Asia, are thought to cure many illnesses; in 1990, China imported 80 tons [72]. A trade study found that in 1994, 44 metric tons of Saiga Horn was exported illegally to China, South Korea, Japan and some European nations. One metric ton is equivalent to 5,000 horns; horn sold for as much as $30 per kilogram in East Asia [39]. In a random survey in August and September 1994, TRAFFIC International investigators found Saiga horn in 131 shops in Hong Kong, from an estimated 15,000 animals. Taiwan banned the sale of Saiga horn in 1994 [39]. Populations of this species have declined in Kazakhstan and Kalmykia and have become endangered in Mongolia. Today, the trade in Saiga horn is so uncontrolled and massive that it threatens the species' future survival [39].

Asian Red Deer (*Cervus elaphus*), known in North America as Elk, have been heavily exploited for their antlers to use in the Traditional Medicine trade, and many races are endangered [65]. A 19th century victim of the Traditional Medicine trade was Schomburgk's Deer (*Cervus schomburgki*). Discovered in eastern Thailand in 1862, no European ever saw the species in the wild [97]. They were heavily hunted for their large and many-tined antlers that supposedly possessed medicinal and magical properties [97]. In the mid-19th century, herds of Schomburgk's Deer were seen in swamps, and during floods, they were pursued by boat, marooned on small islands, and speared [97]. When swamp drainage and irrigation added to their threats, they retreated to bamboo jungles, to which they were not well adapted, until these, too, were cleared for rice fields [97].The last known Schomburgk's Deer was shot by a policeman in September 1932 [97].

The species was considered extinct and officially listed as such by the IUCN [98]. In 1991, a pair of antlers from an unknown type of deer was seen by Laurent Chazee, an agronomist with the United Nations, in a Traditional Medicine shop in a remote part of Laos [73]. Chazee photographed the antlers, which were later identified as coming from a Schomburgk's Deer; the shop owner told him that the animal had been killed the previous year [73].

Forests nearby may shelter more of these deer, and the site is considered by local people to have sacred animal spirits; hunting is prohibited there [73]. A shop in Phnom Penh, Cambodia, in February 1994 offered antlers of what were represented as Schomburgk's Deer for $10 a pair [99]. The seller was obviously unaware of the extraordinary rarity of this deer. There is still no proof that the species survives, and it has been listed as extinct in *2000 IUCN Red List of Threatened Species*.

Bovids are used in many forms of traditional medicine and uncontrolled trade may jeopardize both the long-term survival of some species of bovids and the maintenance of traditional medicine delivery. Although some information on the use of wildlife for medicinal purposes is available from published pharmacopoeias and ethno-biological studies, in most cases little is known regarding harvest and trade volumes, trade controls, market dynamics and conservation impacts [9].

Approximately 80% of the world's population relies on animal and plant based medicines for primary health care [100]. Increased demand and increased human populations are leading to increased and often unsustainable rates of exploitation of wild-sourced ingredients. The high demand for medicinal animals creates additional pressure on natural populations and most importantly on many endangered species that are noted to be in rapid decline. It is therefore imperative to record indigenous knowledge as related to the use of vertebrates and to devise strategies for sustainable utilization of these animals [47].

In this sense, as pointed by Alves et al. [30], the manner in which natural resources are used by human populations and cultural norms associated with that use are extremely relevant to the definition of possible conservation.

It is also important to consider that medicines extracted from animals and plants are significant and valuable resources since they are the unique available remedies for most of the human populations that do not have access to the industrialized drugs and medical care. The socio-cultural aspects are relevant when discussing sustainable development [4, 101]. This social perspective includes the way people become aware of natural resources, their utilization, allocation, transference, and management [102].

Thus, the inclusion of zootherapy in the multidimensionality of the sustainable development is interpreted as a fundamental component to achieve the sustainable use of faunal resources [103]. Soejarto [104] remarks that conservation permits the continuing use of the resources in ways that are non-destructive and sustainable, while from the pharmaceutical point of view, providing time to eventually demonstrate fully the medicinal value of the resources.

There is an urgent need to examine the ecological, cultural, social, and public health implications associated with fauna usage, including a full inventory of the animal species used for medicinal purposes and the socio-cultural context associated with their consumption.

REFERENCES

[1] Alves, RRN; Rosa, IML. Biodiversity, traditional medicine and public health: where do they meet? *Journal of Ethnobiology and Ethnomedicine*, 2007 3(14), 9.

[2] Good, C. Ethno-medical Systems in Africa and the LDCs: Key Issues in Medical Geography. In: Meade MS editor. *Conceptual and Methodological Issues in Medical Geography*. Chapel Hill, NC, USA: University of North Carolina; 1980.

[3] Gesler, WM. Therapeutic landscapes: medical Issues in Light of the new cultural geography. *Social Science and Medicine*, 1992 34(7), 735-746.

[4] Alves, RRN; Rosa, IL. Why study the use of animal products in traditional medicines? *Journal of Ethnobiology and Ethnomedicine*, 2005 1(5), 1-5.

[5] Lev, E. Traditional healing with animals (zootherapy): medieval to present-day Levantine practice. *Journal of Ethnopharmacology*, 2003 85, 107-118.

[6] Mahawar, MM; Jaroli, DP. Traditional knowledge on zootherapeutic uses by the Sahari tribe of Rajasthan, India. *Journal of Ethnobiology and Ethnomedicine*, 2007 3(25), 6.

[7] Mahawar, MM; Jaroli, DP. Traditional zootherapeutic studies in India: a review. *Journal of Ethnobiology and Ethnomedicine*, 2008 4(1), 17.

[8] Mahawar, MM; Jaroli, DP. Animals and their products utilized as medicines by the inhabitants surrounding the Ranthambhore National Park, India. *Journal of Ethnobiology and Ethnomedicine*, 2006 2(46), 5.

[9] Alves, RRN; Rosa, IL; Santana, GG. The Role of Animal-derived Remedies as Complementary Medicine in Brazil. *BioScience*, 2007 57(11), 949-955.

[10] El-Kamali, HH. Folk medicinal use of some animal products in Central Sudan. *Journal of Ethnopharmacology*, 2000 72, 279-282.

[11] Alves, RRN. Fauna used in popular medicine in Northeast Brazil. *Journal of Ethnobiology and Ethnomedicine*, 2009 5(1), 1-30.

[12] Kakati, LN; Ao, B; Doulo, V. Indigenous Knowledge of Zootherapeutic Use of Vertebrate Origin by the Ao Tribe of Nagaland. *Human Ecology*, 2006 19(3), 163-167.

[13] Negi, CS; Palyal, V. Traditional Uses of Animal and Animal Products in Medicine and Rituals by the Shoka Tribes of District Pithoragarh, Uttaranchal, India. *Ethno-Med*, 2007 1(1), 47-54.

[14] Ponting, C. A Green History of the World. 1st ed. London, UK: Sinclair-Stevenson Ltd, 1991.

[15] Gupta, AK. Origin of agriculture and domestication of plants and animals linked to early Holocene climate amelioration. *Current Science*, 2004 87(1), 54-59.

[16] Wyly, E. Urban Origins and Historical Trajectories of Urban Change. *Geography*, 2008(350), 1-10.

[17] Allchin, B; Allchin, R. Origins of a Civilization, The Prehistoric and Early Archaeology of South Asia. 1st ed. New Delhi: Penguin Books, 1997.

[18] MacDonald, GM. Biogeography: Introduction to Space, Time and Life. New York: John Wiley, 2003.

[19] Wilson, DE; Reeder, DM. *Mammal Species of the World. Baltimore*, USA: Johns Hopkins University Press, 2005.

[20] IUCN. IUCN Red List of Threatened Species [online]. 2008 [cited February 2009]. Available from URL: http://www.iucnredlist.org

[21] ITIS. ITIS Catalogue of Life: 2008 Annual Checklist [online]. 2008 [cited February 2009]. Available from URL: http://www.catalogueofife.org/search.php

[22] CITES. CITES Appendix [online]. 2008 [cited February 2009]. Available from URL: http://www.cites.org/eng/resources/species.html

[23] Lev, E. Healing with animals in the Levant from the 10th to the 18th cent. *Journal of Ethnobiology and Ethnomedicine*, 2006 2(11), 9.

[24] Adeola, MO. Importance of wild Animals and their parts in the culture, religious festivals, and traditional medicine, of Nigeria. *Environmental Conservation*, 1992 19(2), 125-134.

[25] Alakbarli, F. Medical Manuscripts of Azerbaijan. Baku, Azerbaijan: Heydar Aliyev Foundation, 2006.

[26] Almeida, CFCBR; Albuquerque, UP. Uso e conservação de plantas e animais medicinais no Estado de Pernambuco (Nordeste do Brasil): Um estudo de caso. *Interciencia*, 2002 27(6), 276-285.

[27] Alves, RRN; Rosa, IL. From cnidarians to mammals: The use of animals as remedies in fishing communities in NE Brazil. *Journal of Ethnopharmacology*, 2006 107, 259–276.

[28] Alves, RRN; Rosa, IL. Zootherapeutic practices among fishing communities in North and Northeast Brazil: A comparison. *Journal of Ethnopharmacology*, 2007 111, 82–103.

[29] Alves, RRN; Rosa, IL. Zootherapy goes to town: The use of animal-based remedies in urban areas of NE and N Brazil. *Journal of Ethnopharmacology*, 2007 113, 541-555.

[30] Alves, RRN; Lima, HN; Tavares, MC; Souto, WMS; Barboza, RRD; Vasconcellos, A. Animal-based remedies as complementary medicines in Santa Cruz do Capibaribe, Brazil. *BMC Complementary and Alternative Medicine*, 2008 8, 44.

[31] Apaza, L; Godoy, R; Wilkie, D; Byron, EH, O; Leonard, WL; Peréz, E; Reyes-García, V; Vadez, V. Markets and the use of wild animals for traditional medicine: a case study among the Tsimane' Amerindians of the Bolivian rain forest. *Journal of Ethnobiology*, 2003 23, 47-64.

[32] Department of the Environment, Water, Heritage and the Arts. Wildlife conservation and complementary medicines [online]. 2003 [cited December 2008]. Available from URL: http://www.environment.gov.au/biodiversity/trade-use/publications/traditional-medicine/index.html

[33] Barboza, RRD; Souto, WMS; Mourão, JS. The use of zootherapeutics in folk veterinary medicine in the district of Cubati, Paraíba State, Brazil. *Journal of Ethnobiology and Ethnomedicine*, 2007 3(32), 14.

[34] Bensky, D; Gamble, A; Kaptchuk, T. Chinese Herbal Medicine Materia Medica Revised Edition. 1st ed. Seattle: Eastland Press, 1993.

[35] Bryan, CP. Ancient Egyptian Medicine. The Papyrus Ebers. 1st ed. Chicago, USA: Ares, 1930.

[36] Cameron, G; Pendry, S; Allan, C; Wu(2004), J. Traditional Asian Medicine Identification Guide for Law Enforcers: Version II. 1st ed. Cambridge, UK: Her Majesty's Customs and excise, London and TRAFFIC International, 2004.

[37] Caprinae Specialist Group. Capricornis sumatraensis [online]. 1996 [cited February 2009]. Available from URL: http://www.iucnredlist.org/search/details.php/3809/all

[38] Carpaneto, GM; Germi, FP. The Mammals in the Zoological Culture of the Mbuti Pygmies in North-Eastern Zaire. *Hystrix*, 1989 1, 1-83.

[39] Chan, S; Madsimuk, AV; Zhirnov., LV. From Steppe to Store: The Trade in Saiga Antelope Horn. 1st ed. Cambridge, UK.: TRAFFIC International, 1995.

[40] CITES. List of animal species used in traditional medicine [online]. 2001 [cited February 2009]. Available from URL: http://www.cites.org/eng/com/AC/17/E17i-05Rev.doc.

[41] CITES. Trade in products, possibly used as medicinals, of the species listed in document AC18 Doc. 13.1 [online]. 2002 [cited December 2008]. Available from URL: http://www.cites.org/common/com/ac/18/E18i-08.doc

[42] CITES. List of species traded for medicinal purposes [online]. 2002 [cited January 2009]. Available from URL: http://www.cites.org/eng/com/ac/18/E18-13-1.pdf

[43] Costa-Neto, EM. Faunistc Resources used as medicines by an Afro-brazilian community from Chapada Diamantina National Park, State of Bahia-Brazil. *Sitientibus*, 1996(15), 211-219.

[44] Costa-Neto, EM. Barata é um santo remédio : introdução à zooterapia popular no estado da Bahia. 1st ed. Feira de Santana, Brazil: EdUEFS, 1999.

[45] Costa-Neto, EM. Recursos animais utilizados na medicina tradicional dos índios Pankararés, que habitam no Nordeste do Estado da Bahia, Brasil. *Actualidades Biologicas*, 1999 21, 69-79.

[46] Costa-Neto, EM; Oliveira, MVM. Cockroach is Good for Asthma: Zootherapeutic Practices in Northeastern Brazil. *Human Ecology Review*, 2000 7(2), 41-51.

[47] Dedeke, GA; Soewu, DA; Lawal, OA; Ola, M. Pilot Survey of Ethnozoological Utilisation of Vertebrates in Southwestern Nigeria, Indilinga. *Afr J Indigenous Knowl Syst*, 2006 5(1), 87-96.

[48] Duckworth, JW; Salter, RE; Khounboline, K. Wildlife in Lao PDR: 1999 Status Report. 1st ed. Gland, Switzerland: IUCN, 1999.

[49] Enqin, Z. Rare Chinese Materia Medica. Shanghai, China: House of Shanghai College of Traditional Chinese Medicine, 1991.

[50] Estes, JW. The Medical Skills of Ancient Egypt. 1st ed: Science History Publications, 1989.

[51] Fischer, F; Linsenmair, KE. Demography of a West African kob (*Kobus kob kob*) population. *African Journal of Ecology*, 2002 40(2), 130-137.

[52] Fitzgerald, S. International wildlife trade: whose business is it? 1st ed. Washington, D.C.: WWF USA, 1989.

[53] Green, MJB. Aspects of the ecology of the Himalayan Musk deer. *Ph.D. thesis*. University of Cambridge, 1985.

[54] Green, MJB. Musk production from Musk deer. In: Hudson RJ, Drew KR, Baskin LM editors. *Wildlife Production System*. Cambridge, UK: Cambridge University Press; 1989.

[55] Guojun, X; Luoshan, X. The Chinese Materia Medica. Beijing: China Medicinal Science and Technology Press, 1996.

[56] Homes, V. On The Scent: Conserving Musk Deer – The Uses of Musk and Europe's Role in its Trade. 1st ed. Brussels, Belgium: TRAFFIC Europe, 1999.

[57] Homes, V. No licence to kill. The population and harvest of musk deer in the Russian Federation and Mongolia. 1st ed. Brussels, Belgium: TRAFFIC Europe, 2004.

[58] Lavigne, D; Wilson, PJ; Smith, RJ; White, BN: Pinniped penises in the marketplace: a progress report., 1999.

[59] Lev, E; Amar, Z. Ethnopharmacological survey of traditional drugs sold in the Kingdom of Jordan. *Journal of Ethnopharmacology*, 2002 82(2-3), 131-145.

[60] Li, S-Y. Habitat management and population change of Hainan Eld's deer. *Chinese Wildlife*, 2000 20(1), 2-3.

[61] Malik, S; Wilson, PJ; Smith, RJ; Lavigne, DM; White, BN. Pinniped Penises in Trade: A Molecular-Genetic Investigation. *Conservation Biology*, 1997 11(6), 1365-1374.

[62] Marques, JGW. Pescando Pescadores: Etnoecologia abrangente no baixo São Francisco Alagoano. 1 st. ed. São Paulo, Brazil: NUPAUB/USP, 1995.

[63] Marshall, NT. Searching for a Cure, Conservation of Medicinal Wildlife Resources in East and Southern Africa. 1st ed. Cambridge, UK: TRAFFIC International, 1998.

[64] Monroy-Vilchis, O; Cabrera, L; Suárez, P; Zarco-González, MM; Soto, CR; Urios, V. Uso tradicional de vertebrados silvestres en la Sierra Nanchititla, México. *INCI*, 2008 33(4), 308-313.

[65] Nilsson, G. The Endangered Species Handbook. 1st ed. Washington, D.C.: The Animal Welfare Institute, 1990.

[66] Nunn, JF. Ancient Egyptian Medicine. 1st ed: University of Oklahoma Press, 1996.

[67] Pieroni, A; Quave, C; Nebel, S; Heinrich, M. Ethnopharmacy of the ethnic Albanians (Arbereshe) of northern Basilicata, Italy. *Fitoterapia*, 2002 73, 217–241.

[68] Powell, AM. Drugs and pharmaceuticals in ancient Mesopotamia. In: I. Jacobs WJ editor. *The Healing Past*. Leiden, Netherlands: Brill; 1993, 47 - 50.

[69] Pujol, J. Natur Africa: The Herbalist Handbook. 1st ed. Durban, S. Africa: Jean Pujol Natural Healers Foundation, 1993.

[70] Rajchal, R. Population Status, Distribution, Management, Threats and Mitigation Measures of Himalayan Musk Deer in Sagarmatha National Park. 1st ed. Babarmahal, Kathmandu, Nepal: DNPWC/TRPAP, 2006.

[71] Ritter, KE. Magical-expert and physician: Notes on two complementary professions in Babylonian medicine. *Assyriological Studies*, 1965 16, 299-323.

[72] Schaller, GB. The Last Panda. 1st ed. Chicago, IL, USA: University of Chicago Press, 1993.

[73] Schroering, GB. Conservation Hotline. Swamp Deer Resurfaces. *Wildlife Conservation*, 1995 98(6).

[74] Sheng, HL. Mustela strigidorsa. In: Sung W editor. *China Red Data Book of Endangered Animals Mammalia*. Beijing, China: Science Press; 1998, 151–152.

[75] Silva, MLVd; Alves, ÂGC; Almeida, AV. A zooterapia no Recife (Pernambuco): uma articulação entre as práticas e a história. *Biotemas*, 2004 17(1), 95-116.

[76] Simelane, TS; Kerley, GIH. Conservation implications of the use of vertebrates by Xhosa traditional healers in South Africa. *South African Journal of Wildlife Research*, 1998 28(4).

[77] Sodeinde, OA; Soewu, DA. Pilot study of the traditional medicine trade in Nigeria. *Traffic Bulletin*, 1999 18(1), 35-40.

[78] Soewu, DA. Wild animals in ethnozoological practices among the Yorubas of southwestern Nigeria and the implications for biodiversity conservation. *African Journal of Agricultural Research*, 2008 3(6), 421-427.

[79] Solanki, GS; Chutia, P. Ethno Zoological and Socio-cultural Aspects of Monpas of Arunachal Pradesh. *Human Ecology*, 2004 15(4), 251-254.

[80] Souza, RF. Medicina e fauna silvestre em Minas Gerais no século XVIII. *Varia Historia*, 2008 24(39), 273-291.

[81] Stetter, C. The Secret Medicine of the Pharaohs - Ancient Egyptian Healing. 1st ed. Chigaco, USA: Edition Q, 1993.
[82] College, JNM. The Chinese Materia Medica Dictionary. 2nd ed. Shanghai, China: Jiangsu New Medical College, 2002.
[83] Thompson, RC. Assyrian medical texts from the originals in the British Museum. 1st ed. Oxford, UK: Oxford University Press, 1923.
[84] Tierra, M. Planetary Herbology, An Integration of Western Herbs Into The Traditional Chines And Ayurvedic Systems. 1st ed. Twin Lakes, Wisconsin, USA: C.A., N.D., O.M.D., Lotus Press, 1988.
[85] True, RH. Folk Materia Medica. *The Journal of American Folklore*, 1901 14(53), 105-114.
[86] Truong, NQ; Sang, NV; Tuong, NX; Son, NT. Evaluation of the wildlife trade in Na Hang District. 1st ed. Ha Noi, Viet Nam: Government of Viet Nam (FPD)/UNOPS/UNDP/Scott Wilson Asia-Pacific Ltd., 2003.
[87] Vázquez, PE; Méndez, RM; Guiascón, ÓGR; Piñera, EJN. Uso medicinal de la fauna silvestre en los Altos de Chiapas, México. *Interciencia*, 2006 31(7), 491-499.
[88] Walston, N. An overview of the use of Cambodia's wild plants and animals in traditional medicine systems. *TRAFFIC Southeast Asia, Indochina*, 2005.
[89] Zhang, B. Musk deer: Their capture, domestication and care according to Chinese experience and methods. *Unasylva*, 1983 35, 16 -24.
[90] Slifkin, RN. Exotic Shofars: Halachic Considerations. 2nd ed: Rosh Hashanah 5768, 2007.
[91] Debbie, NG; Burgess, EA. Against the Grain: Trade in musk deer products in Singapore and Malaysia. 1st ed: WWF/TRAFFIC Southeast Asia, 2005.
[92] Van, NDN; Tap, N. An overview of the use of plants and animals in traditional medicine systems in Viet Nam. 1st ed. Ha Noi, Viet Nam: TRAFFIC Southeast Asia, Greater Mekong Programme, 2008.
[93] Pieroni, A; Giusti, ME; Grazzini, A. Animal remedies in the folk medicinal practices of the Lucca and Pistoia Provinces, Central Italy. In: Fleurentin J, Pelt JM, Mazars G editors. *Des sources du savoir aux médicaments du futur/from the sources of knowledge to the medicines of the future*. Paris: IRD Editions; 2002, 371-375.
[94] Kang, S; Phipps, M. A question of attitude: South Korea's Traditional Medicine Practitioners and Wildlife Conservation. 1st ed. Hong Kong: TRAFFIC East Asia, 2003.

[95] Sheng, H; Ohtaishi, N. The status of deer in China. In: Ohtaishi N, Sheng H-I editors. *Deer of China: Biology and Management*. Amsterdam, The Netherlands: Elsevier Science Publishers; 1993, 1-11.
[96] TRAFFIC. Musk deer [online]. 2002 [cited]. Available from U RL: http://www.undp.org/hdr2003/indicator/indic_4_1_1.html
[97] Day, D. The Doomsday Book of Animals. A Natural History of Vanished Species. 1st ed. New York: Viking Press, 1981.
[98] WCMC (World Conservation Monitoring Centre). 1994 IUCN Red List of Threatened Animals. 1st ed. Gland, Switzerland: International Union for the Conservation of Nature (IUCN), The World Conservation Union, 1993.
[99] Martin, ES; Phipps, M. A Review of the Wild Animals Trade in Cambodia. *TRAFFIC Bulletin*, 1996 16(2), 45–60.
[100] Anon. The Wildlife (Protection) Act, (as amended upto 1991). 1st ed. Dehradun, India: Natraj Publisher, 1993.
[101] Posey, DA. Exploração da biodiversidade e do conhecimento indígena na América Latina: desafios à soberania e à velha ordem. In: CavalcantI C editor. *Meio Ambiente, desenvolvimento sustentável e políticas públicas*. São Paulo: Cortez; 1997, 345–368.
[102] Johannes, RE. Integrating traditional ecological knowledge and management with environmental impact assessment. In: Inglis JT editor. *Traditional Ecological Knowledge: Concepts and Cases*. Ottawa, Canada: International Program on Traditional Ecological Knowledge and International Development Research Centre; 1993, 33–39.
[103] Costa-Neto, EM. Implications and applications of folk zootherapy in the State of Bahia, Northeastern Brazil. *Sustainable Development*, 2004 12, 161–174.
[104] Soejarto, DD. Biodiversity prospecting and benefit-sharing: perspectives from the field. *Journal of Ethnopharmacology*, 1996 51, 1–15.

In: New Developments in Biodiversity … ISBN: 978-1-61324-374-9
Editor: Thomas W. Pace

Chapter 3

CONSERVATION AND MANAGEMENT OF THE BIODIVERSITY IN A HOTSPOT CHARACTERIZED BY SHORT RANGE ENDEMISM AND RARITY: THE CHALLENGE OF NEW CALEDONIA

Roseli Pellens ***and Philippe Grandcolas***
UMR 7205 CNRS, Département Systématique et Evolution,
Muséum national d'Histoire naturelle,
45, rue Buffon, 75005 Paris, France

ABSTRACT

New Caledonia is a peculiar hotspot, a small-sized island (ca. 17,000 km²), relatively isolated from any continent (ca. 1200 km from Australia), with moderately high mountains and complex orography. Its biota is very rich in endemic species and highly endangered. Our analysis of the number of references in systematics, ecology and conservation shows that its biota attracts attention of scientists since long ago. In a first and long period, references focused on the description of the biodiversity. More recently these descriptions were intensified and complemented by ecological and later by conservation studies. Recent researches on phylogenetics and biogeography indicated that the biota of New

* E-mail: pellens@mnhn.fr and pg@mnhn.fr.

Caledonia is characterized by short range endemism and rarity, with three patterns of endemism: (I) species regionally endemic to New Caledonia distributed in the whole range of an ecosystem; (II) short range endemics with parapatric/allopatric distributions; (II) short range endemics with disjunct distributions. Researches on conservation showed that this biota is highly endangered due to three main threats: fire, mining and invasive species. In this chapter we detail these threats and elaborate a model based on their frequency and spatial distribution to understand how they could affect species with contrasting patterns of endemicity. Our analysis show that the conservation of the biodiversity in a context where species are dominantly short range endemics and rare is a main problem to be faced by New Caledonian authorities as well as by scientific researchers that must provide the basis for political decisions. Regardless the biogeographical pattern of endemism the chances of survival of rare species with short ranges in the case of large scale habitat destruction are quite low. In the case of threats that are more restricted in area, the loss of a species disjunctly distributed is more problematic in terms of loss of phylogenetic diversity. In this case, speciation by niche conservatism can be hypothesized to be less frequent, thus each species can be implied to be more original and in stronger need of conservation by itself. In addition, due to the distances from one another, the number of closely related species in a small island can be much lower than in the case of species with parapatric/allopatric distribution. Fires, mining and introduced species need special control. The two firsts for the habitat destruction they promote over extensive areas and the later by the possibility of continuing to endanger even in areas officially protected.

INTRODUCTION

The biota of New Caledonia is very rich in endemic species and highly endangered. Therefore, the region has been classified as a hotspot of biodiversity (Myers et al. 2000). Hotspots are by definition these regions with conflicts of interest between conservation of amazingly rich and diverse biota and deleterious effects of man occupancy. We will review briefly the main characteristics of New Caledonia geography and biodiversity and then we will examine how they can be affected by the main local man-induced ecosystem disturbances.

New Caledonia is a peculiar hotspot indeed, a small-sized island with moderately high mountains (peaking at ca. 1600 m) and complex orography, and relatively isolated from any continent (ca. 17,000 km^2; 1200 km from Australia). But it is also a very ancient piece of land and for this reason it has often been seen as a continental island and a Gondwanan refuge because of the occurrence of several relict taxa (e.g., *Amborella*, the endemic sister-group of all other flowering plants) and of a very old geological basement (80 My). In this context, high richness and endemism have been interpreted as the result of a long-term evolution in isolation. Recent biogeographical studies have revised this view, by considering both detailed geological studies and phylogenetic relationships of endemic taxa (Murienne et al., 2005; Grandcolas et al., 2008). The island could not have conserved in situ its relicts because this piece of Gondwana that collided with an island arc at the limit of two tectonic plates was submerged until 37 My ago. Accordingly, regionally endemic groups often resulted from several recent dispersal founding events since 37 My. Local short range endemism is also extremely important and is even necessarily more recent, at least partly dating back to quaternary climatic variations (e.g., Murienne et al., 2008a).

This new biogeographical paradigm came out with the recent reviving of evolutionary and ecological studies in New Caledonia. Until a few years ago, studies dealing with New Caledonian biota were mostly taxonomic studies or inventories produced at a rising pace (Figure 1a, b). This trend of an ever-increasing taxonomic production can also be detected with the examination of bibliographies (O'Reilly, 1955; Pisier, 1983; Chazeau, 1995) or by looking at the description series of the very rich local fauna (Tillier, 1988; Chazeau and Tillier, 1991; Matile et al., 1993; Najt andMatile , 1997; Najt and Grandcolas, 2002; Grandcolas, 2008a, 2009).

In lag with this trend, the number of evolutionary and ecological studies increased only recently (Figure 1a, b), showing that the background natural history knowledge brought back by taxonomic studies and inventories began to be fruitfully used further than for classification issues (Grandcolas, 2008b). This growing interest for understanding the origin and the functioning of New Caledonian biota took place in the context of higher threats to the local ecosystems that also determined a more recent increase in studies of conservation biology and ecosystem management (Figure 1a, b).

Looking at these three lagged trends (Figure 1a, b), one can identify three steps in the study of New Caledonian biodiversity, by adding successively to the description of the biota, the understanding of processes and then the management of the ecosystems and conservation issues.

As a matter of consequence, we are at a very critical moment for facing the threats to the ecosystem (Beauvais et al., 2006; Pascal et al., 2008), having now different cards in our hands to understand the situation and to propose some cautions or some remedies.

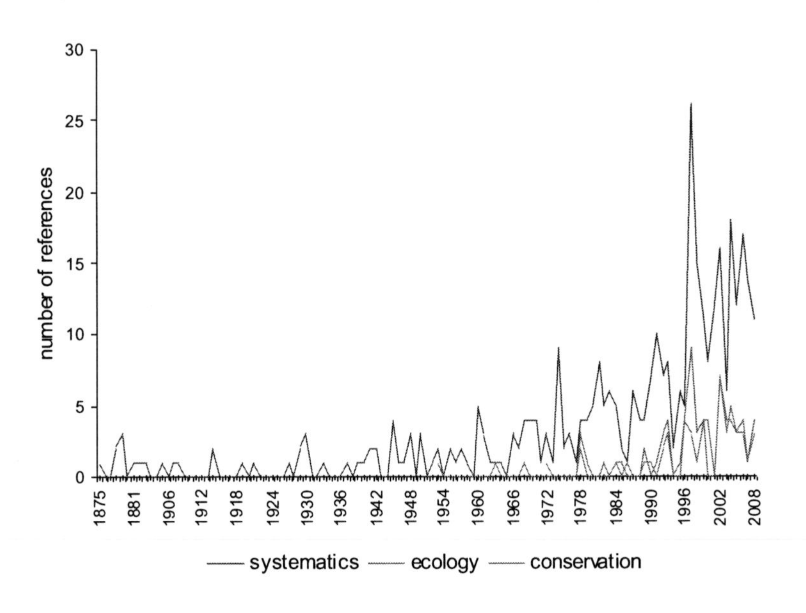

Figure 1a. Number of references dealing with systematics, ecology and conservation in New Caledonia, published from 1875 to 2008. Data from Zoological Records. The search for "systematics" was made with the word New Caledonia in the title "and" the combined results for the words new species, systematic*, taxonom*, phylogen*, evolut*, diversity and inventory in the topics. The search for "ecology" was made with the word New Caledonia in the title "and" the combined results for the words ecology, population, community, ecosystem and process in the topics. The search for "conservation" was made with the word New Caledonia in the title "and" the combined results for the words conservation, threat, invasive, introduction and extinction in the topics. In this search we considered references dealing with present terrestrial biota including fresh water, and excluding marine and paleontological data.

In this perspective, the biodiversity of New Caledonia can be characterized with the following characteristics. First, patterns of short range endemism appear more and more frequent as distributional studies accumulate (Haase and Bouchet, 1998; Bradford and Jaffré, 2004; Pellens, 2004; Murienne et al., 2005, 2008a; Munzinger et al., 2008). In this prevalent case, sister species apparently appeared by allopatric speciation are often separated by very short distances. Such speciation has often taken place by niche conservatism without a strong adaptive differentiation and many short range endemics are therefore quite similar to each other. Conversely, some species look like relicts, with disjunct distributions in so-called refuge areas (e.g., Pintaud et al., 2001).

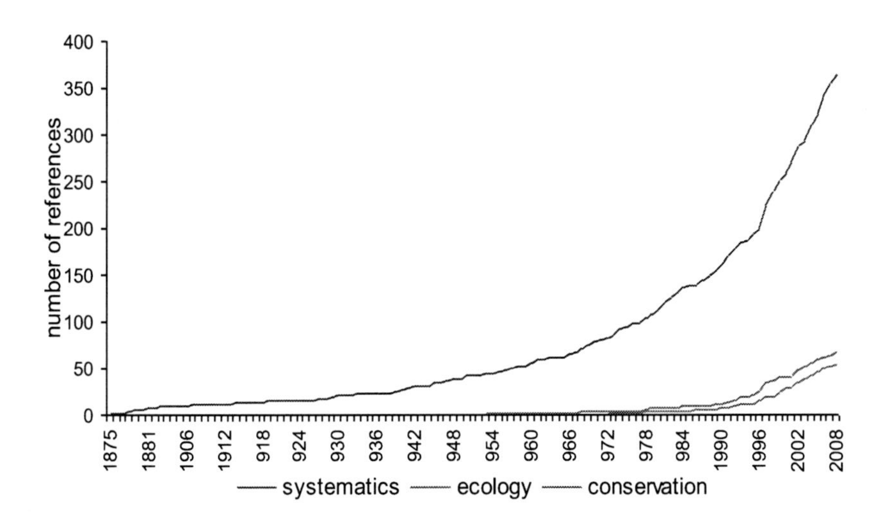

Figure 1b. Accumulation curves of the references dealing with systematics, ecology and conservation in New Caledonia, from 1875 to 2008. Data from Zoological Records (see details in the legend of Figure 1a).

When such a differentiation occurred however, adaptation to the high local soil diversity including metalliferous soils is important, together with adaptation to climatic and orographic diversity. Short range endemism is not the only distinctive character of the local biota, since species rarity – population low density or high patchiness – is also commonplace. Both endemism and rarity are expectedly correlated as a result of several scale effects or specific mechanisms (Gaston, 1994). Consequently, many New Caledonian species have very small population size.

The recent degradation of vegetation under the action of man adds one more biological constraint to the ecosystems, by fragmentation of the main forest types into large but dissected massifs of evergreen forest and into small and overdispersed pieces of sclerophyll forests, separated by cultivated areas, pastures or poor savannas.

We will successively detail the main threats to the ecosystems, and then we will infer the potential and specific impact of these threats, given the characteristics of New Caledonian biodiversity. Finally, we will recommend future directions for the study and conservation of local biota.

MAIN THREATS TO THE NEW CALEDONIAN BIODIVERSITY

During the last two decades several studies/symposiums/meetings/reports were made in order to characterize the main dangers to the biodiversity in New Caledonia, and to elaborate recommendations to the conservation and management of natural resources. Although most of their recommendations took long to be put in practice, these studies were unanimous in calling attention to three major threats that work exclusively or in combination to put in danger the native biota: fire, mining, and introduced species. These threats along with insufficient number of protected areas, inadequate laws concerning introduced species, and developmental policies of economic help contrary of the maintenance of the local biodiversity contribute to destroy terrestrial, lotic and lagoon ecosystems, putting in danger a great number of species (Bouchet et al., 1995; Gabrié et al., 1995; Gargominy et al., 1996; Jaffré et al., 1998; Ekstrom et al., 2000; Gargominy, 2003; Beauvais et al., 2006). Here we will first make a brief review of these three threats for further exploring their impacts on endemic species.

Fires

The destruction of forests by fire is the most ancient and up-to-date problem in the New Caledonian mainland, being the major causes of habitat loss in the lowlands. Traditionally, fires were used to clean pastures and cultivated fields before a new season of plantation, as a cultural practice. Nowadays it is often used as a cultural habit repeated every year in the same places without any immediate reason, i.e., even when there is not agriculture or pasture. It is also lit by hunters in grasslands and open savannas to enhance the growth of new grass, or to facilitate the access (Bouchet et al., 1995; Gabrié et al., 1995). The main outcome of these activities is that every year during the dry season uncontrolled fires destroy thousands of ha of savannas, arboreous savannas, sclerophyll forests (also called dry forests) and shrublands (locally called *maquis* or *maquis minier* when talking about the vegetation on the areas disturbed by mining), which constitute the most sensitive ecosystems. To have an idea of its importance, 10 024 ha were destroyed in 1995, 3 340 in 1996, 21 678 in 1997, 2 558 in 1998, 5075 in 1999 and 15 710 in 2000, i.e., a surface of 3.5% of the mainland burned in six years

(Gargominy, 2003). The ecosystem most affected by fire is sclerophyll forests. Almost totally destroyed, it is now reduced to a few fragments that contain about than 2% of its original surface (Bouchet et al., 1995).

Mining

New Caledonia is exceptionally rich in rare metals, and presently the extraction of nickel (plus chromium and cobalt, although in minor quantities) constitute the base of its economic activity as well as its most striking environmental disturbance (Gabrié et al., 1995). Although the surface exploited is very small (0.9% of the mainland surface), the mining impacts spread over very large surfaces, totally changing the landscape of the mountains, and also impacting the rivers and the lagoon, as well as the areas near the roads used to transport the minerals. In the mountains, the opencast mineral exploitation creates a reddish landscape marked by the presence of openings, access roads, screes and enormous areas of soil erosion. In addition to that, another problem is that this kind of landscape degradation is produced in several mountains, very often promoted by small sites of prospecting, hardly ever leading to the exploitation. It is useless to say that the first impact of the activities associated to mining is the elimination of the native vegetation, provoking immediate habitat loss, which certainly leads to local extinction of species whose ranges of distribution coincide with the areas impacted. Considering that the main vegetation of areas used for mineral exploitation is the shrublands "*le maquis*" with 92% of endemic species, the biodiversity loss caused by this activity can be enormous (Gabrié et al., 1995; Dupon, 1996; Gargominy, 2003; Pascal et al., 2008).

Introduced Species

As an insular ecosystem, New Caledonia attracts attention to the importance of the introduction of exotic species, especially because some of these species became invasive and provoked serious damages to native biota and ecosystems. This way, presently we have an extensive inventory of the introduced species in the archipelago (McKee, 1994; Gargominy et al., 1996; Beauvais et al., 2006) and some results permitting to evaluate their impact on the native biota (Gargominy et al., 1996; Le Breton et al., 2005; Jourdan, 1997; 2006; Jourdan and Mille, 2006; Meyer et al., 2006; Pascal et al., 2006).

The number of introduced species in New Caledonia is rather high, having increased markedly during the last 50-60 years. Recent estimations indicate 1,412 species of plants (64 invasive, with nine of them in the IUCN list of "100 World's Worst Invasive Alien Species") (McKee, 1994; Gargominy et al., 1996; Meyer et al., 2006); 518 of invertebrates (six in IUCN the list of "100 World's Worst Invasive Alien Species") (Gargominy et al., 1996; Le Breton et al., 2005; Jourdan, 2006; Jourdan and Mille, 2006); and 42 of vertebrates (12 in the IUCN list of "100 World's Worst Invasive Alien Species") (Gargominy et al., 1996; Pascal et al., 2006). Although most of the introduced species remain in the surroundings of man, some invasive species spread their population through wild ecosystems, putting in danger wild populations of native species.

One case well studied concerns the invasive fire ant *Wasmannia auropunctata*, considered "the most dangerous pest ever introduced in this archipelago" (Jourdan, 1997), being a main threat for human population as well as to the New Caledonian autochthon biodiversity. In the wild, they were shown to outcompete local ant species leading to their disappearance in the sites invaded (Jourdan, 1997; Le Breton et al., 2005); to cause the reduction of the diversity and density of lizards, especially geckos (Bauer and Sadlier, 1993); and to endanger several ha of Niaouli savannah due to the disruption of the ecological balance by reducing predators and/or competitors of a mealybug that produces a honeydew used by a fungus with black spores that makes a black film on the plant's leaves, ultimately resulting in the obstruction of the photosynthesis (Cochereau and Potiaroa, 1995).

Among introduced vertebrates reported to have a severe effect in New Caledonia's native biota there are pigs, goats, cattle and deers. Individuals of these species graze the understory of humid and dry forests, trampling on young shoots preventing plants from regenerating, compacting the soil, reducing the litter mass, and changing microclimate at ground level (Pascal et al., 2006). Their activities strongly change the habitats, making difficult forest regeneration, and reducing the populations of understory organisms, as shown for species of the mollusc *Placostylus* (Brescia et al., 2008). Considering the rarity of many plant species in New Caledonian biota, the action of invasive vertebrates can be striking giving the final touch to the species extinction, as shown by Bouchet et al. (1995). In fact, species with very rare or very localized populations are extremely fragile, in a way that any action on their habitat lead to deterministic extinctions. Among plants there are several examples of species that are invading or becoming a danger.

A recent evaluation of the subject made by Meyer et al. (2006) attracts attention to *Pinus caribea* largely used for reforestation in areas with acid soils and on ultramafic soils, now considered the only allochtone species able to develop in the ultramafic soils (Morat et al., 1999), becoming a potential danger to the species endemic of the *maquis* as well as to the vegetation of other canopy-opened ecosystems.

CHALLENGES FOR CONSERVATION OF HORT RANGE ENDEMIC AND RARE SPECIES

Short range endemism and rarity are the key characteristics of New Caledonian biota (Grandcolas et al., 2008). Among endemic species one can find basically three types. The first is the case of species regionally endemic to New Caledonia, distributed in the whole range of an ecosystem, as for example, *Araucaria montana,* one of the endemic species of *Araucaria* from New Caledonia that occurs in the *maquis* and in the dense and humid forests from middle and high altitude (Veillon, 1980) (Pattern I, Table 1). This kind of distribution is not necessarily the most frequent. The second type is represented by short range endemic species with *parapatric* distributions, i.e., closely related species having contiguous distribution ranges, and *allopatric* distribution, i.e., with species separated by very short distances. In this last case, it is common to observe close related species separated by distances not greater than 5 km, with or without a clear geographical barrier, as is the case of species of the endemic cricket genus *Agnotecous* (Desutter-Grandcolas and Robillard, 2006), or the endemic cockroach genus *Angustonicus* (Pellens, 2004; Murienne et al., 2005) and *Lauraesilpha* (Murienne et al., 2008a,b). We assume that closely related species with these two patterns of geographic distribution are affected in a similar way by most threats, thus we consider them as a single case (Pattern II, Table 1). Short range endemics with parapatric/allopatric distribution are often inferred to have arisen by speciation with niche conservatism (Wiens, 2004; Murienne et al., 2008b, 2009). In this case, species maintain most of their characteristics and niche dimensions (like the specialization to particular habitats, or to particular forms of exploring habitat or food resources), in way that closely related species share most of their phylogenetic characteristics, and one species can represent better the diversity of its whole group. This aspect has important implications for conservation.

Table 1. Patterns of endemism and types of geographic distribution of the threats in New Caledonia. Pattern I: species regionally endemic to New Caledonia, distributed in the whole range of an ecosystem; Pattern II: related species with parapatric/allopatric distributions; Pattern III: related species with disjunct distributions. Type A: concentrated threat distributed in the whole range of an ecosystem; Type B: sparse threat distributed in the whole range of an ecosystem; Type C: concentrated threat limited to a small area/region; Type D: sparse threats limited to a small area/region. The frequency and the effect associated to the extension and distribution of the threat in related species of each pattern of endemism are evaluated as + low; ++ medium; +++ medium/high; ++++ high

Table 7.1. Overview of Variables in the Multinominal Logit Model

Dependent Variable	Variable name	Categories		Freq.	Perc.	Scale
	LANDUSE	Forest		42,717	77.02%	Pixel
		Annual crops		4,165	7.51%	Pixel
		Perennial crops		4,225	7.62%	Pixel
		Grassland		4,357	7.86%	Pixel
Explanatory Variables	**Variable name**	**Mean**	**Std. Dev.**	**Minimum**	**Maximum**	**Scale**
Geophysical variables						
Slope (degrees)	SLOPE	10.79	8.03	0	45	pixel
Elevation (100 m)	ELEV	10.49	3.70	0.25	23.51	pixel
Aspect (degrees)	ASP	187	105	0	360	pixel
Aspect squared (degrees)	ASP^2	45,967	38,822	0	129,598	pixel
Distance to river (100 m)	TORIVER	1.15	1.11	0	9.60	pixel
Location variables						
Distance to road (100 m)	TOROAD	84.29	72.42	0	322.74	pixel
Distance to hamlet (km)	TOVILL	3.47	2.95	0	20.28	pixel
Dummy inside the National Park	PARK	0.33	0.47	0	1	pixel
Distance to city (km)	TOCITY	92	33	9	145	hamlet
Distance to city squared (km)	$TOCITY^2$	9,527	5,937	77	21,091	hamlet
Distance to district capital (km)	TODISTRICT	27	19	0	67	hamlet
Socioeconomic variables						
# years irrigation infrastructure	IRR_YRS	3.80	9.53	0	71	village
ln (population)	lnPOP	6.64	0.67	5.46	8.45	village
% of population in working age	POPWORK	67	8	39	83	village
Spatially lagged variables						
Lagged slope (degrees)	LAG	10.76	6.85	0	35	pixel
Lagged slope * slope	SLOPE_LAG	211.49	222.92	0	1,680	pixel

The third pattern of short range endemism often observed in New Caledonia is represented by related species with *disjunct distributions*, i.e., closely related species with short ranges of distribution but found in distant areas, often in relict or refuge habitats, as the case of the palms studied by Pintaud et al. (2001) for example (Pattern III, Table 1). In this case, speciation by niche conservatism can be hypothesized to be less frequent, thus each species can be implied to carry more autapomorphies. In addition, due to the distances from one another, the number of closely related species in a small island like New Caledonia can be much lower than in the case of species with parapatric/allopatric distributions. These aspects indicate that in a clade with species disjunctly distributed, the loss of a species is more problematical in terms of loss of phylogenetic diversity, than in the case of species with parapatric/allopatric distributions.

Independently of their pattern of distribution, species are most often very rare. Species distributed in wide areas (Pattern I) are often rare due to low population density or restricted microhabitats (even if this microhabitat is widely distributed). Short range endemics are rare due to the restricted areas of distribution, which makes them difficult to find, regardless the population density. Nevertheless, the most common pattern observed in New Caledonia is even more extreme: rare species with narrow ranges of distribution and low population densities, which makes very small population sizes. In Table 1 we present an evaluation of the frequency and the effect of the spatial distribution of the main threats to the New Caledonian biota on the three types of endemic species.

This evaluation is based on a model combining the three kinds of endemicity and the distribution and the intensity of the threats. Concentrated threats widespread all over the territory or all over a type of ecosystem (Type A) are the rarest ones. But their effects are the most dangerous being able to eliminate wide range species or several related short range species, independently of their pattern of endemism. In spite of the uncommonness of this kind of event, we have the concrete example of the destruction of the sclerophyll forests, practically totally destroyed by fires that burned here and there every dry season during several decades, in which most of the species are highly endangered (Bouchet et al., 1995). Conversely, impacts scarcely distributed, but widespread over the territory or over an ecosystem (Type B) are the most frequent (like fires at the edges of humid forests, the effects of several invasive species in different parts of the ecosystems, forest clearing for small plantations, hunting, selective logging, or the combination of these activities). The main effect of this kind of impact is the reduction of the population size of several species at the same time period. It can be of moderate consequences on species with Pattern I of endemism, but could lead to the extinction of at least one species with Patterns II and III, or highly endanger several related species. Nevertheless, due to the lower number of species and the higher phylogenetic information carried by each species, the impacts of this kind of threat is higher in species with disjunct distribution.

Threats that are very concentrated and limited to a small area/region (Type C) are of moderate frequency. Mining is a good example of this kind of threat. Since this activity changes the landscape of a mountain totally eliminating the vegetation and destroying the natural habitats, it is prone to lead one or several related species with parapatric/allopatric distribution to the extinction (Pattern II), due to the short scale of distribution of short range endemics and the short distances between different related species, and one with disjunct distribution. In opposition, its effects on species with Pattern I are moderate, i.e., it can lead to the reduction of population size, but individuals can survive elsewhere, due to the spatial restriction of the threat. Once more, due to lower number of species and higher phylogenetic information, species with disjunct distribution are the most affected.

Finally, threats that are sparse and limited to a small area/region (Type D) are of moderate to high frequency. This type of effect can be illustrated by invasive species as well as hunting, or selective logging. These activities usually occur in several places in a given area and are repeated several times. Their effects can be of minor importance on wide range species (Pattern I), due to their large scale of distribution, and moderate consequences on close related species with Pattern II, due to the fact that there will be some species that survive. Nevertheless, the impact on species with Pattern III of short range endemism can be strong, since they are usually very few related species in New Caledonia, therefore loosing a lot with the extinction of a single species.

CONCLUSION

In conclusion, the conservation of the biodiversity in a context where species are dominantly short range endemics and rare is a main problem to be faced by New Caledonian authorities as well as by scientific researchers that must provide the basis for political decisions in this respect.

By considering three types of endemic species facing different kinds of threats, we showed that the biodiversity of New Caledonia which comprises a remarkably high number of short range endemics (Patterns II and III) is especially vulnerable to the most common threats. In most cases, small populations of short range endemics will be severely affected. In the long term, small and declining populations have few chances of lasting several years or decades, even if they can survive some time in very small untouched or less disturbed patches. This situation is even more problematical when one considers that population densities are often very low, making remnant populations especially small. In this context, fires, mining and introduced species deserve special attention and control. The two firsts for the habitat destruction they promote over extensive areas and their repetition in space all over the territory, that makes them prone to lead several related species to be extinguished in a short time period. The later by the possibility of continuing to endanger even in areas officially protected.

Phylogenetic and genetic studies must also be developed in order to better understand the corollaries of the extinctions of some short range endemics, in terms of conservation of the biodiversity of New Caledonia whose evolutionary patrimonial value is invaluable.

ACKNOWLEDGMENT

The present chapter has been made within the framework of the project ANR BIONEOCAL (grant ANR Biodiversité 2007, Philippe Grandcolas). We thank Dr. Tony Robillard and Dr. Eric Guilbert for kindly accepting reviewing this chapter and giving suggestions that contributed to make it clearer.

REFERENCES

Bauer, A.M. and Sadlier, R.A. (1993) Systematics, biogeography and conservation of the lizards of New Caledonia. *Biodiversity Letters*, 1, 107-122.

Beauvais, M.-L., Coléno, A. and Jourdan, H. (Eds.) (2006) *Les espèces envahissantes dans l'archipel néo-calédonien - Un risque environnemental et économique majeur*. IRD editions, Paris, 259p + CD.

Bouchet, P., Jaffré, T. and Veillon, J.-M. (1995) Plant extinction in New Caledonia: protection of sclerophyll forest urgently needed. *Biodiversity and Conservation*, 4, 415-428.

Bradford, J. and Jaffré, T. (2004) Plant species microendemism and conservation of montane maquis in New Caledonia: two new species of *Pancheria* (Cunoniaceae) from the Roche Ouaïème. *Biodiversity and Conservation*, 13, 2253-2274.

Brescia, F., Pollabauer, C.M., Potter, M.A. and Robertson, A.W. (2008) A review of the ecology and conservation of Placostylus (Mollusca: Gastropoda: Bulimulidae) in New Caledonia. *Molluscan Research*, 28, 111-122.

Chazeau, J. (1995) *Bibliographie indexée de la faune terrestre de Nouvelle-Calédonie. Systématique, écologie et biogéographie.* ORSTOM Editions, Paris, 95 p.

Chazeau, J. and Tillier, S. (Eds.) (1991) Zoologia Neocaledonica 2. *Mémoires du Muséum national d'Histoire naturelle*, 149, 1-358.

Cochereau, P. and Potiaroa, T. (1995) Caféiculture et *Wasmannia auropunctata* (Hymenoptera, Formicidae, Myrmicinae) en Nouvelle-Calédonie, ORSTOM, Nouméa 20p.

Desutter-Grandcolas, L. and Robillard, T. (2006) Phylogenetic systematics and evolution of *Agnotecous* in New Caledonia (Orthoptera: Grylloidea, Eneopteridae). *Systematic Entomology*, 31, 65-92.

Dupon, J.E. (1986) *The effects of mining on the environment of high islands: a case study of nickel mining in New Caledonia.* South Pacific Regional Environment Programme, Nouméa, 6 p.

Ekstrom, J.M.M., Jones, J.P.G., Willis, J. and Isherwood, I. (2000) *The humid forests of New Caledonia: biological research and conservation recommendations for the vertebrate fauna of Grande Terre.* CSB Conservation Publications, 100 p.

Gabrié, C., Licari, M.L. and Mertens, D. (1995) *L'état de l'environnement dans les Territoires Français du Pacifique Sud : La Nouvelle Calédonie*, L'institut Français de l'environement, Paris, 115 p.

Gargominy, O. (2003) *Biodiversité et conservation dans les collectivités françaises d'autre mer*. Paris, Comité Français pour l'UICN, 246+X p.

Gargominy, O., Bouchet, P., Pascal, M., Jaffré, T. and Tourneur, J.-C. (1996) Conséquences des introductions d'espèces animales et végétales sur la biodiversité en Nouvelle-Calédonie. *Revue d'Ecologie (Terre Vie)*, 51, 375-402.

Gaston, K.J. (1994) *Rarity*. Chapman and Hall, London, 205 p.

Grandcolas, P. (Ed.) (2008) Zoologia Neocaledonica 6. Biodiversity studies in New Caledonia. Paris, *Mémoires du Muséum national d'Histoire naturelle*, 197, 1-326.

Grandcolas, P. (2008b) Introduction. *In:* Grandcolas, P. (Ed.) Zoologia Neocaledonica 6. Biodiversity studies in New Caledonia. *Mémoires du Muséum national d'Histoire naturelle, Paris*, 197, 9-12.

Grandcolas, P. (Ed.) (2009). Zoologia Neocaledonica 7. Biodiversity studies in New Caledonia. *Mémoires du Muséum national d'Histoire naturelle,* 198, in press.

Grandcolas, P., Murienne, J., Robillard, T., Desutter-Grandcolas, L., Jourdan, H., Guilbert, E. and Deharveng, L. (2008) New Caledonia: a very old Darwinian island? *Philosophical Transactions of the Royal Society of London, B*, 363, 3309-3317.

Haase, M. and Bouchet, P. (1998) Radiation of crenobiontic gastropods on an ancient continental island: the *Hemistomia*-clade in New Caledonia (Gastropoda: Hydrobiidae). *Hydrobiologia*, 367, 43–129.

Jaffré, T., Bouchet, P. and Veillon, J.-M. (1998) Threatened plants of New Caledonia: is the system of protected areas adequate? *Biodiversity and Conservation*, 7, 109-135.

Jourdan, H. (1997) Threats on Pacific islands: the spread of the Tramp Ant *Wasmannia auropunctata* (Hymenoptera: Formicidae). *Pacific Conservation Biology*, 3, 61-64.

Jourdan, H. (2006) Les invertébrés menaçants pour l'archipel néo-calédonien : recommandations pour leur prévention. *In:* Beauvais, M.-L., Coléno, A. and Jourdan, H. (Eds.) *Les espèces envahissantes dans l'archipel néo-calédonien - Un risque environnemental et économique majeur*. IRD editions, Paris, pp. CD 215-245.

Jourdan, H. and Mille, C. (2006) Les invertébrés introduits dans l'archipel néo-calédonien : espèces envahissantes et potentiellement envahissantes. Première évaluation et recommandations pour leur gestion. *In:* Beauvais, M.-L., Coléno, A. and Jourdan, H. (Eds.) *Les espèces envahissantes dans l'archipel néo-calédonien - Un risque environnemental et économique majeur*. IRD editions, Paris, pp. CD 163-214.

Le Breton, J., Jourdan, H., Chazeau, J., Orivel, J. and Dejean, A. (2005) Niche opportunity and ant invasion: the case of Wasmannia auropunctata in a New Caledonian rain forest. *Journal of Tropical Ecology*, 21, 93-98.

Loope, L.L. and Pascal, M. (2006) Quelques espèces animales envahissantes aux frontières de la Nouvelle-Calédonie et présentant un risque environnemental majeur. *In:* Beauvais, M.-L., Coléno, A. and Jourdan, H. (Eds.) *Les espèces envahissantes dans l'archipel néo-calédonien - Un risque environnemental et économique majeur*. IRD editions, Paris, pp. CD 246-257.

Matile, L., Najt, J. and Tillier, S. (Eds.) (1993) Zoologia Neocaledonica 3. *Mémoires du Muséum national d'Histoire naturelle*, 157, 1-218.

McKee, H.S. (1994) *Catalogue des plantes introduites et cultivées en Nouvelle-Calédonie*. Muséum national d'Histoire naturelle, Paris, 164 pp.

Meyer, J.-Y., Loope, L.L., Sheppard, A., Munzinger, J. and Jaffré, T. (2006) Les plantes envahissantes et potentiellement envahissantes dans l'archipel néo-calédonien : première évaluation et recommandations de gestion. *In:* Beauvais, M.-L., Coléno, A. and Jourdan, H. (Eds.) *Les espèces envahissantes dans l'archipel néo-calédonien - Un risque environnemental et économique majeur*. IRD editions, Paris, pp. CD 50-115.

Morat, P., Jaffré, T. and Veillon, J.-M. (1999) Menaces sur les taxons rares et endémiques de la Nouvelle-Calédonie. *Bulletin de la Société Botanique du Centre-Ouest (SBCO),* 19, 129-144.

Morat, P., Jaffré, T. and Veillon, J.-M. (2001) The flora of New Caledonia's calcareous substrates. *Adansonia*, 23, 109-207.

Munzinger, J., McPherson, G. and Lowry, P.P. (2008) A second species in the endemic New Caledonian genus *Gastrolepis* (Stemonuraceae) and its implications for the conservation status of high-altitude maquis vegetation: coherent application of the IUCN Red List criteria is urgently needed in New Caledonia. *Botanical Journal of the Linnean Society*, 157, 776-783.

Murienne, J., Grandcolas, P., Piulachs, M. D., Bellés, X., D'Haese, C., Legendre, F., Pellens, R. and Guilbert, E. (2005) Evolution on a shaky piece of Gondwana: is local endemism recent in New Caledonia? *Cladistics* 21, 2-7.

Murienne, J., Pellens, R., Budinoff, R.B., Wheeler, W. and Grandcolas, P. (2008a) Phylogenetic analysis of the endemic New Caledonian cockroach *Lauraesilpha*. Testing competing hypothesis of diversification. *Cladistics*, 24, 802-812.

Murienne, J., Pellens, R. and Grandcolas, P. (2008b) Short range endemism in the cockroach Lauraesilpha (Blattidae, Tryonicinae) in New Caledonia: distribution and new species. In: Grandcolas, P. (Ed.) Zoologia Neocaledonica 6, Systematics and Biodiversity in New Caledonia. *Mémoires du Muséum National d'Histoire Naturelle*, 197, 261-271.

Murienne, J., Guilbert, E. and Grandcolas, P. (2009) Species diversity in the New Caledonian endemic genera Cephalidiosus and Nobarnus (Insecta: Heteroptera: Tingidae), an approach using phylogeny and species distribution modeling. *Biological Journal of the Linnean Society*, 95, in press.

Myers, N., Mittermeier, R.A., Mittermeier, C.G., Fonseca, G.A.B. and Kent, J. (2000) Biodiversity hotspots for conservation priorities. *Nature*, 403, 853-858.

Najt, J. and Matile, L. (Eds.) (1997) Zoologia Neocaledonica 4. *Mémoires du Muséum national d'Histoire naturelle*, 171, 1-399.

Najt, J. and Grandcolas, P. (Eds.) (2002) Zoologia Neocaledonica 5. Systématique et endémisme en Nouvelle-Calédonie., *Mémoires du Muséum national d'Histoire naturelle*, 187, 1- 283.

O'Reilly, P. (1955) *Bibliographie méthodique, analytique et critique de la Nouvelle-Calédonie*. Societé des Océanistes, Paris, 361 p.

Pascal, M., Barré, M., Garine-Wichatitsky, M., Lorvelec, O., Frétey, T. and Brescia, F. (2006) Les peuplements néo-calédoniens de vertébrés : invasions, disparitions. *In:* Beauvais, M.-L., Coléno, A. and Jourdan, H. (Eds.) *Les espèces envahissantes dans l'archipel néo-calédonien - Un risque environnemental et économique majeur.* IRD editions, Paris, pp. CD 111-162.

Pascal, M., Richer de Forges, B., Le Guyader, H. and Simberloff, D. (2008) Mining and other threats to the New Caledonia biodiversity hotspot. *Conservation Biology*, 22, 498-499.

Pellens, R. (2004) Nouvelles espèces d'*Angustonicus* (Insecta, Dictyoptera, Blattaria, Tryonicinae) et endémisme du genre en Nouvelle-Calédonie. *Zoosystema*, 26, 307-314.

Pintaud, J.-C., Jaffré, T. and Puig, H. (2001) Chorology of New Caledonian palms and possible evidence of Pleistocene rain forest refugia. *Comptes rendus de l'Académie des Sciences de Paris, Sciences de la vie*, 324, 453-463.

Pisier, G. (1983) Bibliographie méthodique, analytique et critique de la Nouvelle-Calédonie. 1955-1982. *Publications de la Société d'Etudes Historiques de la Nouvelle-Calédonie*, 34, 1-350.

Tillier, S. (1988) (Ed.) Zoologia Neocaledonica 1. *Mémoires du Muséum national d'Histoire naturelle*, 142, 1-158.

Veillon, J.-M. (1980) Architecture des espèces néo-calédoniennes du genre Araucaria. *Candollea*, 35, 609-640.

Wiens, J.J. (2004) Speciation and ecology revisited: phylogenetic niche conservatism and the origin of species. *Evolution*, 58, 193-197.

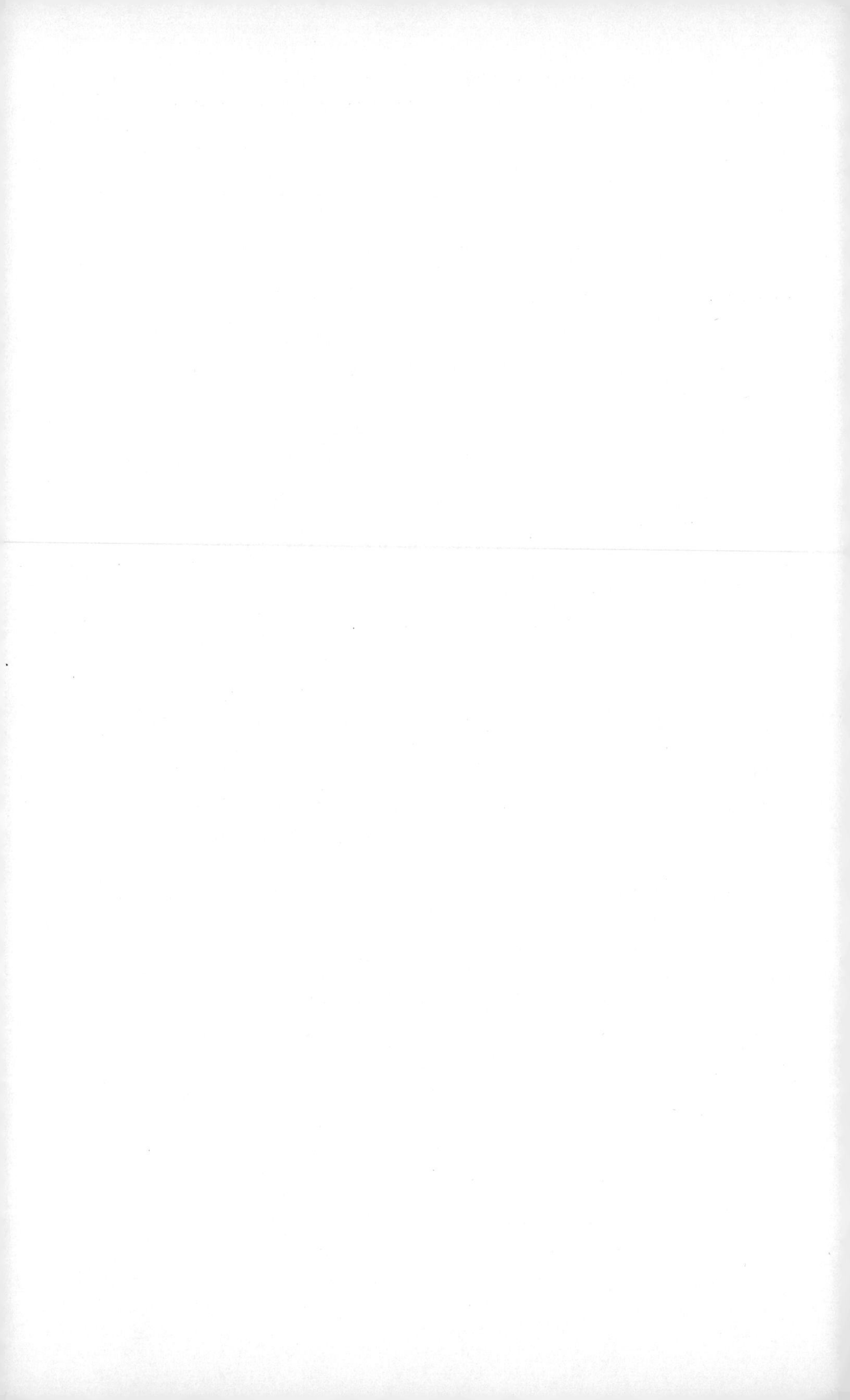

In: New Developments in Biodiversity ... ISBN: 978-1-61324-374-9
Editor: Thomas W. Pace

Chapter 4

Surface Stratification of Soil Nutrients in no Till Limits Nutrient Availability and Reduces Grain Yield

***B. J. Radford**[*1] **and B. A. Cowie**[2]*

[1]Queensland Department of Environment and Resource Management, LMB 1, Biloela, 4715 Queensland, Australia

[2]Queensland Department of Environment and Resource Management, PO Box 1762, Rockhampton, 4700 Queensland, Australia

Abstract

Research has shown that the less mobile elements in the soil, such as P, K and Zn, become stratified in the surface 50 mm of soil after several years of continuous no-till cropping. This causes nutritional constraints to productivity when the surface soil becomes dry unless the sub-surface soil is highly fertile or receives appropriate levels of fertiliser.

No-till crops grown in the semi-arid, subtropical environment of central Queensland, Australia, are particularly prone to nutritional disorders as a result of surface stratification. The surface soil can remain dry for long periods during crop growth while the roots obtain water stored in the heavy clay soils during the preceding fallow. Consequently nutrient deficiency symptoms have appeared despite apparently

---

[*] Email: Bruce.Radford@derm.qld.gov.au

satisfactory levels of those same nutrient elements in the 0-100 mm layer, which is the standard sampling depth used to assess soil fertility.

The original aim of the 13-year experiment reported here was to assess the effect of tillage frequency and intensity, and stubble retention and removal, on soil water storage, soil nutrient status, and the growth and yield of rainfed grain sorghum in central Queensland. During the first 7 years of research, the mean yields of no till and traditional tillage (disc plough and scarifier) without fertiliser application were not significantly different. As the experiment continued, it became apparent that supplementary nutrition was required. The placement of appropriate levels of fertiliser elements (P, K and Zn) 100 mm deep and 50 mm to the side of each row resulted in large yield responses to no till. The yield of no till in comparison with traditional tillage (both with stubble retained and appropriate fertiliser application) was 147%, 129% and 174% in the final three crops, respectively. These responses also reflect the outstanding potential of longer-term no till that has been demonstrated in other research work in central Queensland. Since mixing of the surface soil (by tillage) gave similar responses, it was concluded that some crop nutrients had become concentrated in the dry surface soil and were therefore inaccessible to plant roots during dry periods.

Keywords: no till, grain yield, surface stratification, soil nutrients

INTRODUCTION

A number of studies have found that the less mobile elements in the soil, such as P, K and Zn, become concentrated in the surface 50 mm of soil after several years of continuous no till cropping [Robson & Taylor, 1987; Cowie *et al.*, 1996]. This causes nutritional constraints to productivity when the surface soil becomes dry unless the rest of the soil profile is highly fertile or receives appropriate levels of fertiliser.

Crops grown in the semi-arid, subtropical environment of central Queensland, Australia, are particularly prone to nutritional disorders as a result of surface stratification. The surface soil can remain dry for long periods during crop growth while the roots obtain water stored in the heavy clay soils during the preceding fallow. This has resulted in nutrient deficiency symptoms after only 5-7 years of no till [Cowie *et al.*, 1996] despite apparently satisfactory levels of those same nutrient elements in the 0-100 mm layer, which is the standard sampling depth used to assess soil fertility. The high yield responses associated with no till in central Queensland [Radford &

Thornton, 2009] encourage the practice of continuous no till and hence the accumulation of nutrients in the surface soil, as well as increasing the rate of nutrient export. The nutrient most susceptible to stratification is phosphorus [Wright *et al.*, 2007].

The experiment reported here was planned at the start of the tillage revolution in Australia [Triplett & Dick, 2008]. Its original aim was to assess the effect of tillage frequency and intensity and stubble retention and removal on soil water storage, soil nutrient status, and the growth and yield of rainfed grain sorghum. The first seven years of research were reported in detail in the scientific literature [Thomas *et al.*, 1990 a, b; Standley *et al.*, 1990]. The mean yields of no till, blade plough/rod weeder tillage and disc plough/scarifier tillage were not significantly different during those first seven years when no fertiliser was applied. As the experiment continued, it became apparent that supplementary nutrients were required. When the appropriate elements (N, P, K, S and Zn) were applied at appropriate rates, there were large positive yield responses in the no-till treatment. A single tillage operation in the no-till treatment prior to sowing in order to mix the surface soil and redistribute the soil nutrients also gave a large positive yield response.

MATERIALS AND METHODS

Site

The experiment was located on a property at Mt Murchison near Biloela (24.32°S, 150.53°E) in eastern central Queensland, Australia. The land originally supported a forest dominated by brigalow (*Acacia harpophylla*) before it was cleared between 1930 and1937. It is known to have been cropped annually since 1962, and no fertiliser was applied until December 1985.

The soil type is a grey Vertisol classified as an Entic Pellustert [Soil Survey Staff, 1975]. The clay content increases down the profile from 41% (0-100 mm) to 53% (140-160 mm). Details of some soil properties prior to the start of the experiment are given in Thomas *et al.* [1990a]. Land slope is about 3%.

The mean annual rainfall is 675 mm (range 317-1132 mm), with 72% falling in summer (October-March). The mean annual evaporation from a class A pan is 1870 mm.

Design and Treatments

The experimental design was originally a randomised block with 6 treatments and 4 replicates. The 6 treatments were disc, blade and no tillage (D, B and N) x stubble retained and stubble removed (+ and -). Disc and blade tillage were carried out as primary tillage operations, and a scarifier was used for all secondary tillage. Disc/scarifier tillage was the conventional fallow management technique when the experiment began in 1978. Blade/rod weeder tillage is a form of reduced tillage designed to reduce soil disturbance and stubble incorporation. In the no till treatment, all weed control was carried out using herbicides, and the only soil disturbance was caused by the sowing tines. Row spacing for the sorghum crops was 1 m except in 1990 when it was 0.75 m. The cotton was sown in twin rows 1 m apart with 3 m between the twins. The total number of tillage operations in both the disc and blade treatments was 47, an average of about four per fallow. The total number of spraying operations in the no-till treatment was 51, also an average of about 4 per fallow. Plots measured 25 x 15 m, and small diversion banks were constructed along the sides of the plots to prevent water running off one plot to the next. Water from the diversion banks drained into grassed laneways at the ends of the plots. After 7 sorghum crops, the plots were split for fertiliser application (control and fertilised), and a further 5 crops (4 sorghum and 1 cotton) were grown. The same tillage x stubble treatments were always maintained on the same plots, and the same fertiliser treatments on the same split plots, each year.

Table 1 shows the details of the cropping sequence, the amounts of fallow and in-crop rainfall received and the amounts of fertiliser elements applied to the plots that were split for fertiliser application.

In 1991, the final year of the experiment, the unfertilised subplots in 3 reps were split into 20 smaller subplots to apply various combinations of nutrient elements and two soil inversion treatments. The aim was to determine the level of response to five nutrient elements and to mixing of the surface soil to overcome surface stratification. Nutrients were applied in the following amounts: N (75 kg/ha), P (40 kg/ha), K (50 kg/ha), Zn (5 kg/ha) and S (30 kg/ha). This fertiliser was banded 100 mm deep and 50 mm to the side of each row. Soil inversion was done before sowing with a single pass of a Rotavator set to a depth of 120 mm. The 20 fertiliser and inversion treatments are listed in Table 2. The design incorporated single, double and triple omission of the five nutrient elements in order to determine what was limiting when all other nutrients were in ample supply.

Table 1. Details of the crops grown, rainfall received and fertiliser applied to the split plots

Crop and year of harvest	Cultivar	Sowing date	Harvest date	Fallow		Crop		Fertiliser applied to split plots
				Period days	Rain-fall mm	Period days	Rain-fall mm	kg/ha
Sorghum 79	E57+	21.2.79	27.7.79	250	598	156	129	Nil
Sorghum 80	E57+	16.1.80	29.4.80	173	274	104	109	Nil
Sorghum 81	E57+	9.1.81	11.5.81	255	231	132	366	Nil
Sorghum 82	E57+	30.12.81	29.4.82	233	293	121	266	Nil
Sorghum 83	E57+	19.1.83	19.5.83	265	277	120	543	Nil
Sorghum 84	E57+	11.1.84	3.5.84	237	336	113	128	Nil
Sorghum 85	E57+	19.12.84	2.5.85	230	421	134	377	Nil
Sorghum 86	E57+	19.12.85	16.4.86	231	462	118	241	50N + 10P
Sorghum 87	E57+	16.12.86	7.5.87	244	470	142	324	38N+10P+ 37K
Sorghum 88	E57+	9.2.88	10.6.88	278	378	122	320	75N + 40P + 82K + 39S
Cotton 90	Siokra	1.11.89	12.3.90	509	921	131	127	24N + 44P + 82K + 42S + 5Zn
Sorghum 91	DK 44M	18.1.91	8.5.91	312	579	110	76	75N + 40P + 50K + 30S + 5Zn

Measurements

Soil samples to assess surface stratification were taken on 25 January 1991 from only the disc and no till treatments, each with stubble retained and without any history of applied fertiliser. Depth increments sampled were 0-25, 25-50, 50-100, 100-150 and 150-200 mm. The 0-25 mm sample was taken using a square-section hand trowel, and the deeper samples were taken through this depression using a 25 mm diameter coring tube with a machined tip. For both treatments, 10 samples of each depth increment were obtained from three replicates, and the 30 samples were bulked and thoroughly mixed. Subsamples were then taken for analysis for bicarbonate-extractable P, exchangeable K, DTPA-extractable Zn and pH by the methods of Rayment & Higginson [1992].

Table 2. Combinations of nutrient element and inversion treatments in the 1991 sorghum crop

Treatment	Groupings
Control	Unfertilised control
S	S only
SN (no P, K or Zn) SP (no N, K or Zn) SK (no N, P or Zn) SZn (no N, P or K)	Triple nutrient omission treatments (selected)
SNP (no K or Zn) SNK (no P or Zn) SNZn (no P or K) SPK (no N or Zn) SPZn (no N or K) SKZn (no N or P)	Double nutrient omission treatments (selected)
SNPK (no Zn) SNPZn (no K) SNKZn (no P) SPKZn (no N) NPKZn (no S)	Single nutrient omission treatments
SNPKZn	All 5 nutrients
Soil inversion Inversion + SNPKZn	Inversion treatments

Soil macrofauna were counted in all plots in May 1990 using a spade sampling technique [Robertson & Simpson, 1988] to collect 10 samples (180 x 180 mm) to a depth of 150 mm in each plot. All soil fauna exceeding ~0.5 mm in length were recorded. Samples therefore included some larger mesofauna [Swift *et al.*, 1979].

Grain yields were measured by means of a small plot header and, in 1990, with a mechanical cotton picker. In 1991, the smaller plots were hand-harvested. Grain moisture contents of subsamples were measured with a moisture meter, and all yields were adjusted to 12% moisture content. Cotton lint yield was measured at ambient moisture content. The datum area was 4 x 20 m in each plot except for the cotton in 1990 (4 x 10 m) and the sorghum in 1991 (1.5 x 3 m). The datum area was reduced in 1991 because in this final year all plots were further subdivided to impose multiple tillage x fertiliser treatments [Asghar *et al.*, 1993].

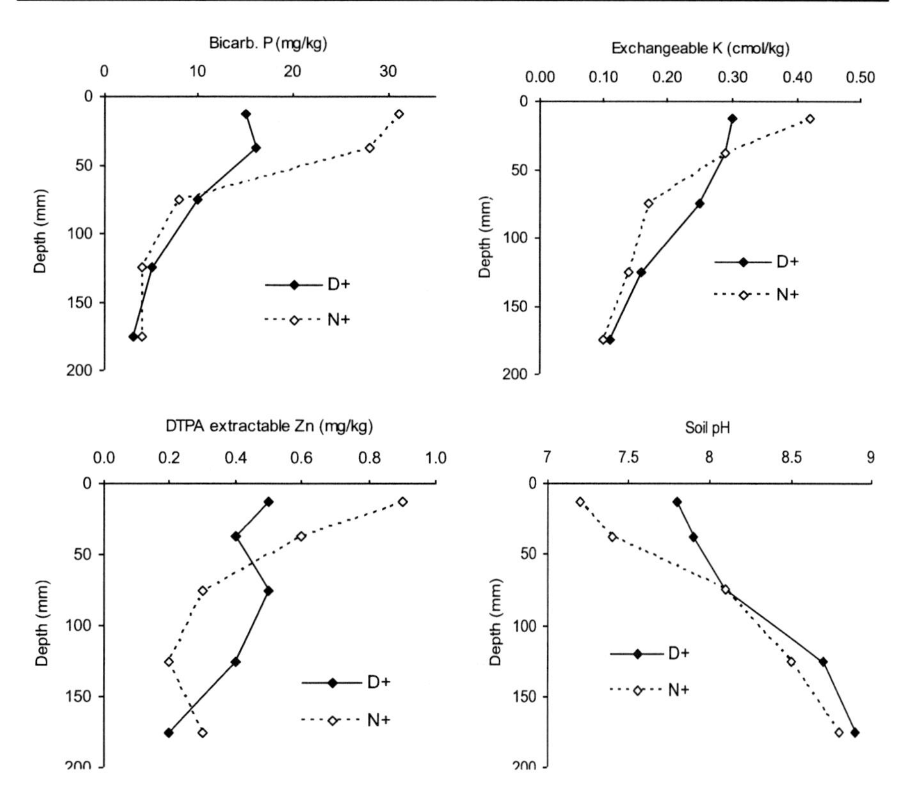

Figure 1. Effect of disc/scarifier tillage (D) and no till (N), both with stubble retained (+) on bicarbonate-extractable P, exchangeable K, DTPA-extractable Zn and soil pH in the surface soil in 1991.

RESULTS

Nutrient Levels in the Surface Soil

The concentration of bicarbonate-extractable P was higher in N+ than D+ in the surface soil (0-25 and 25-50 mm) but levels did not vary with tillage practice at greater depths (50-200 mm) (Figure 1). Exchangeable K was higher in N+ at 0-25 mm but correspondingly lower in N+ at 50-100 mm. Extractable Zn was higher in N+ at 0-25 and 25-50 mm but lower in N+ at 50-100 and 100-150 mm. Soil pH was lower in N+ than D+ in the surface soil (0-25 and 25-50 mm) while values at greater depths were similar. These changes in the properties of the surface soil under no till resulted in marked deficiency symptoms in the cotton crop after 12 years of treatment application (Figure 2).

Deficiency symptoms were also observed in the blade tillage treatments but to a lesser degree.

Soil Macrofauna

The population density of macrofauna in the soil after 12 years was significantly (P<0.01) higher in N+ than any other treatment. There were 82 m^{-2} in N+, 8 in N-, 20 in B+, 11 in B-, 26 in D+ and 11 in D-. Isopods comprised 46% of the population in N+.

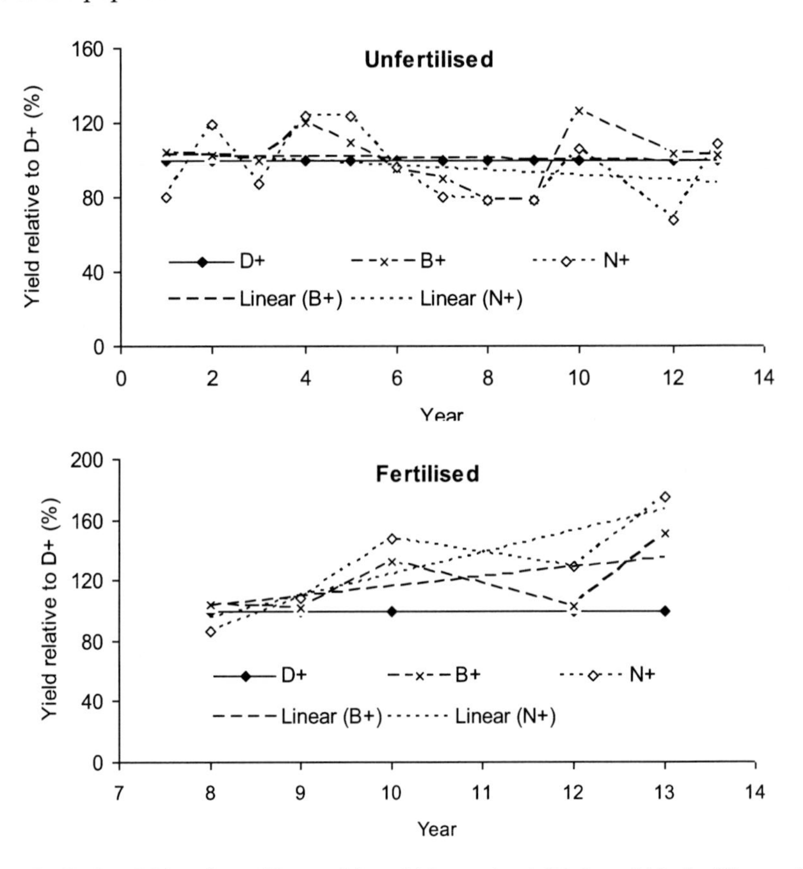

Figure 3. Grain yields of no tillage with stubble retained (N+) and blade tillage with stubble retained (B+) relative to traditional disc tillage with stubble retained (D+) for 12 unfertilised crops (top) and 5 fertilised crops.

Figure 2. Cotton crop at Mt Murchison after 12 years of no till with stubble removed showing nutrient deficiency symptoms due to surface stratification (foreground) and the response to applied fertiliser (N, P, K, S and Zn).

Grain Yields

During the 13 years of the experiment, there was little difference in the mean yields of D+, B+ and N+ without applied fertiliser (Table 3; Figure 3). No till yielded slightly less than the tilled treatments. In contrast, when adequate amounts of the appropriate nutrient elements were applied to the final three crops, N+ greatly outyielded D+ (Table 4; Figure 3). The mean yield of N+ in the final three crops was half as much again as D+ (150%) while B+ yielded a quarter as much again (129%). These dramatic responses contrast sharply with the non-response without fertiliser, and indicate large yield gains are possible in continuous no till when appropriate crop nutrition is provided. N+ yielded 3.44 t/ha of sorghum in 1988 (D+: 2.34 t/ha) and 3.48 t/ha in 1991 (D+: 1.99 t/ha). N+ produced 511 kg/ha of cotton in 1990 while D+ gave only 396 kg/ha.

It should be emphasised that these responses to N+ occurred only when stubble was retained. When stubble was removed at the end of each season, the yield from no till was reduced, both with and without fertiliser application (Figure 4).

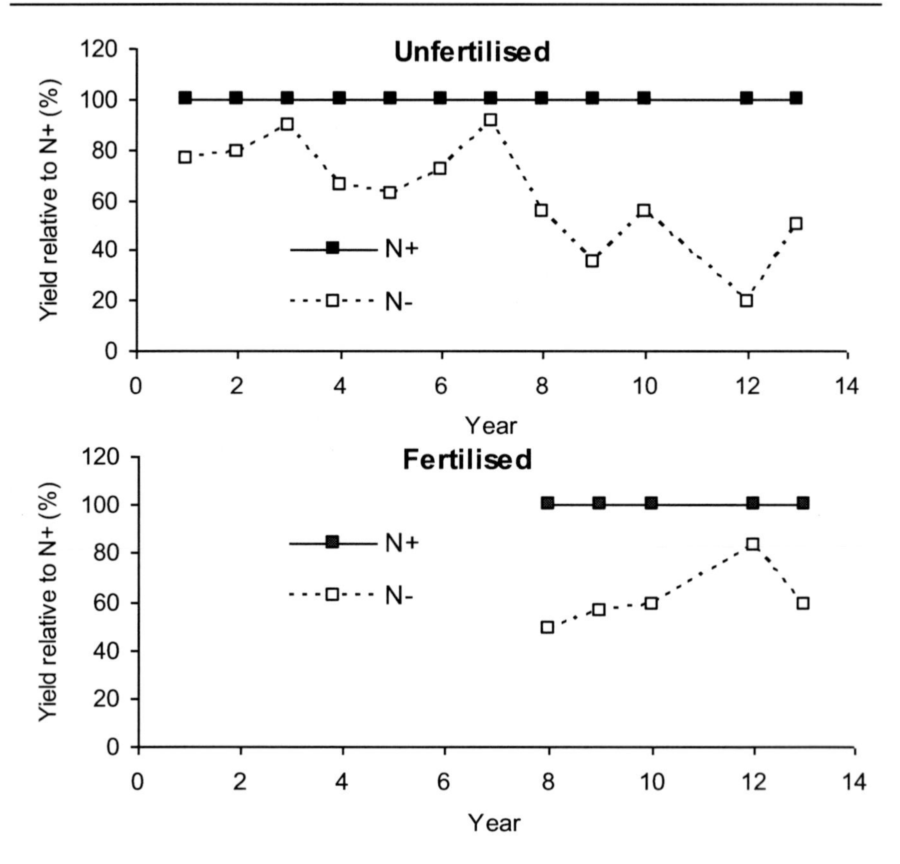

Figure 4. Effect of stubble retention (+) and removal (-) on crop yield under a no-till (N) regime with and without fertiliser application.

Yield responses to the different combinations of nutrient elements and to soil inversion are shown in Table 5. There was no significant (P>0.05) yield increase in response to the addition of any single nutrient. The addition of P+K+Zn, however, always significantly (P<0.05) outyielded the control. If we compare the fully fertilised treatment with the five treatments lacking a single nutrient, yield was reduced only when P was removed. Where two elements were missing, yield was reduced by excluding P+N, P+K or P+Zn (i.e. all three treatments with P omitted). Where three elements were missing, yield reductions again occurred only when one of the excluded elements was P. Soil inversion resulted in a yield not significantly different from the complete fertiliser treatment.

Table 3. Statistical significance of treatment differences in grain yield (t/ha) for 12 crops without applied fertiliser. Treatments are disc tillage (D), blade tillage (B) and no till (N) with stubble retained (+) and stubble removed (-). na: not available

Treatment	Crop number												Means
	1	2	3	4	5	6	7	8	9	10	11	12	
D -	3.32ab	1.40c	3.08bc	2.82cd	2.03c	2.65a	4.52abc	2.52b	1.30b	1.93b	0.22	1.09b	2.24
D+	3.61a	1.65b	3.59a	2.95bcd	2.26bc	2.97a	5.09a	3.00a	1.95a	2.13b	0.30	2.10a	2.63
B -	3.71a	1.68b	3.29ab	3.07abc	2.04c	2.77a	4.99ab	2.40b	1.61a	2.14b	0.25	2.44a	2.53
B+	3.74a	1.68b	3.57a	3.53ab	2.48b	2.83a	4.58abc	2.35b	1.52ab	2.69a	0.31	2.15a	2.62
N -	2.20c	1.55bc	2.77c	2.41d	1.74d	2.06b	3.71c	1.19c	0.51c	1.25c	0.04	1.14b	1.71
N+	2.90b	1.96a	3.12bc	3.64a	2.79a	2.86a	4.08bc	2.16b	1.43b	2.26b	0.21	2.27a	2.47
lsd (P=0.05)	0.60	0.25	0.41	0.64	0.28	0.45	0.92	0.45	0.45	0.36	na	1.04	na

Table 4. Statistical significance of treatment differences in grain yield (t/ha) for five crops with applied fertiliser. Treatments are disc tillage (D), blade tillage (B) and no till (N) with stubble retained (+) and stubble removed (-). na: not available

Treatment	Crop number					Means
	8	9	10	11	12	
D -	2.43c	1.55b	2.26b	0.37	2.24b	1.77
D+	3.10ab	2.07a	2.34b	0.40	1.99b	1.98
B -	2.50c	2.04a	2.31b	0.41	2.42b	1.94
B+	3.23a	2.09a	3.10a	0.41	2.99ab	2.36
N -	1.31d	1.25b	2.05b	0.42	2.07b	1.42
N+	2.69bc	2.24a	3.44a	0.51	3.48a	2.47
lsd (P=0.05)	0.45	0.45	0.37	na	1.04	na

Table 5. Grain yield responses of 1991 sorghum to fertiliser and soil inversion treatments in no till with stubble retained (N+)

Grain yields in order of magnitude*	Treatment
3867 a	Inversion + SNPKZn
3623 ab	SPK
3475 abc	SNPKZn
3469 abc	SPKZn
3420 abc	NPKZn
2949 abcd	Inversion
2824 bcde	SNP
2747 bcde	SPZn
2665 bcde	SNPZn
2526 cdef	SP
2469 cdef	SNPK
2273 def	Control
2217 def	SKZn
2202 def	SK
2102 def	SNKZn
2082 def	SN
1959 def	SZn
1907 ef	S
1821 ef	SNZn
1531 f	SNK

* Values followed by the same letter are not significantly different (P>0.05).

DISCUSSION

Stratification of relatively immobile soil nutrients (such as P, K, Zn, Cu, Mn, Fe and B) has been previously reported in response to no till [Robson & Taylor, 1987]. There were no associated yield reductions due to P, K or Zn stratification, however, probably because the soils were highly fertile, highly fertilised or moist enough at the surface to enable root access. Stratification problems appeared on brigalow soils in central Queensland after 7 years of no till at Mt Murchison and after only 5 years at Brigalow Research Station [Cowie *et al.*, 1996]. At both sites the surface soil (0-50 mm) was generally too dry for nutrient extraction by crop roots and levels of plant-available nutrients below 50 mm were low. At such sites the traditional 0-100 mm soil sample is inappropriate for assessing fertility because nutrients found at 0-50 mm are largely unavailable. A 50-150 mm sample would be more useful, as suggested by Cowie *et al.* [1996].

At a third site in central Queensland (Biloela), however, no stratification occurred after 7 years of no till [Cowie *et al.*, 1996]. Here soil mixing may have occurred during the passage of the sowing tines, when surface soil fell into soil cracks, and when earthworms and termites were active. The sowing tines were 1 m apart at Mt Murchison but 0.275 m apart at Biloela and Brigalow during the three years prior to sampling. Soil cracking would have been similar at all three sites because the clay content of their surface soil (0-100 mm) was similar (25-41%). At Biloela high earthworm population densities in no till (17 m^{-2}) [Wilson-Rummenie *et al.*, 1999] may have aided in redistributing the immobile elements [Darwin, 1881]. Termite activity under no till [Wilson-Rummenie *et al.*, 1999] may also have reversed stratification at Biloela.

Termites were also present at Mt Murchison [Holt *et al.*, 1993] but failed to overcome the stratification problem. A high total macrofauna population in N+ (86 m^{-2}) also failed to have any appreciable effect. This high population density of soil macrofauna, however, is a sign of superior soil health in N+ and may have enhanced soil structure, rainfall infiltration and water use efficiency.

Apart from the stratification problem, which can be overcome by adding nutrients, the condition of the N+ soil must have improved markedly after 10 years to produce the mean yield response reported here (150% of D+). Possible reasons include:

- Improved surface soil structure
- Reduced soil compaction

- Increased biological activity
- Increased soil organic matter and
- Increased plant-available water holding capacity through the leaching of salt below the crop root zone.

Tillage causes deterioration of soil physical properties due to the pulverising and smearing effects of implements, and pressure from load-bearing wheels, resulting in a marked reduction in infiltration capacity [Clarke, 1987]. Complete repair of the damage done by tillage implements, tractors and harvesting machinery takes several years. For example, soil compaction damage has persisted for 5 years under no till management [Radford *et al.*, 2007].

While compaction damage remains, soil macrofauna populations will also be low [Radford *et al.*, 2001] and will therefore contribute little to soil structural improvement. It has been shown that soil macrofauna take about 7 years to become abundant and diverse in temperate Australia after cultivation has ceased [Douglas, 1987].

Organic carbon at Mt Murchison declined at a slower rate in N+ than the other five treatments during the first 7 years of the experiment [Standley *et al.*, 1990]. This indicates that differences may have become larger after 10-13 years (when yield differences also became larger).

Before the experiment commenced, the levels of soil chloride below a depth of one metre were high: 811 mg kg^{-1} at 1.0-1.2 m, 991 at 1.2-1.4 m and 1090 at 1.4-1.6 m [Thomas *et al.*, 1990a]. Rayment & Bruce [1984] classed soil Cl levels of 600-2000 mg/kg as high. There were no notable changes during the first 3-5 years of the experiment [Standley *et al.*, 1990] but leaching may have occurred in subsequent years as rainfall infiltration into N+ improved. Any reduction in the soil Cl levels at 1.0-1.6 m depth would have increased the plant-available water content of the soil by allowing increased water extraction by plant roots [Dang *et al.*, 2008].

Our large responses to long-term (>10 years) no till have been duplicated in another experiment in central Queensland [Radford & Thornton, 2009]. Research in northern China has shown that continuous long-term (15 years) no till improved soil structure and biological activity [Li *et al.*, 2007]. This included significantly increased aggregate stability in the larger size classes, and greater capillary porosity (pores of diameter <60 μm). Canadian research showed that 20 years of continuous no till with full stubble retention but no nitrogen application outyielded 2 years of no-till; the 20-year no till produced 2.8 t/ha and the 2-year no-till 1.8 t/ha [Derpsch, 2008]. Increase in soil carbon content was considered the main factor leading to this result. Benefits from no till accrue over time, and Sá [2004] has recognised 4 phases in the evolution of a long-term no-till system: initial (0-5 years), transition (5-10 years), consolidation (10-20 years) and maintenance (>20 years). It is only in the maintenance phase that maximum soil benefits (physical, chemical and biological) are achieved.

Derpsch [2008] claims that any tillage performed during the earlier phases of no till means a return to the initial phase. If this is true, tillage to invert the soil and make surface-stranded nutrients available to crop roots merely fixes one problem and creates other longer-lasting ones. Although tillage makes nutrients in the surface soil more available, it also disrupts the network of vesicular arbuscular mycorrhizhae in no till [Kabir, 2004], which enhances nutrient uptake from the deeper soil.

This experiment has now been terminated but archived soil and plant samples can be made available to interested researchers for further analyses.

CONCLUSION

Surface stratification of nutrient elements with low soil mobility may occur after several years of continuous no-till management. Concentration of these elements in the surface soil can result in crop nutrient deficiencies when the surface soil is dry because the elements cannot be taken up by the plant roots. As a result, marked yield reductions may occur. The elements implicated were P, K and Zn. The problem can be overcome by applying the deficient elements in fertiliser banded 100 mm deep and 50 mm to the side of each row. It can also be overcome with a single tillage operation to invert the soil, but tillage can do more harm than good in the long term. When crop nutrition was adequate, long-term (>10 years) no till with stubble retained gave large yield responses compared with traditional tillage with stubble retained. Yield increased by 47%, 29% and 74% in the final three crops (years 10, 12 and 13, respectively).

ACKNOWLEDGMENT

We thank Greg Thomas, Adrian Webb, John Standley and co-workers for instigating this work; Greg Thomas and Don Sinclair for leading the 1986 and

1987 experiments; Heather Hunter for leading the 1988 experiment; Les Robertson and his team for conducting the soil macrofauna counts; the late Mohammad Asghar for his small-plot work in the 1991 sorghum to ascertain the nutrients responsible for the yield response to no till; and Gary Blight, Melindie Hastie, Stephen Hunt, Mal Hunter and David Lack for their valuable assistance.

REFERENCES

[1] Asghar, M., Lack, D. W., Cowie, B. A. and Parker, J. C. (1993). Effects of surface soil mixing after long-term disc, blade and zero tillage on sorghum production in central Queensland. In: G. K. McDonald & W. D. Bellotti (Eds.), *Proceedings of the 7th Australian Agronomy Conference* (300-303). Parkville, Australia: The Australian Society of Agronomy, Inc.

[2] Clarke, A. L. (1987). The need for fertility restoration/maintenance. In: J. P. Thompson & J. A. Doughton (Eds.), *Proceedings of the Queensland Crop Production Conference 1987* (7-17). Brisbane, Australia: QDPI Conference and Workshop Series QC 87007.

[3] Cowie, B. A., Hastie, M., Hunt, S. B., Asghar, M. and Lack, D. W. (1996). Surface soil nutrient distribution following zero tillage and traditional tillage management. In: M. Asghar (Ed.), *Proceedings of the 8th Australian Agronomy Conference* (160-163). Toowoomba, Australia: The Australian Society of Agronomy, Inc.

[4] Dang, Y. P., Dalal, R. C., Mayer, D. G., McDonald, M., Routley, R., Schwenke, G. D., Buck, S. R., Daniells, I. G., Singh, D. K., Manning, W. and Ferguson, N. (2008). High subsoil chloride reduces soil water extraction and crop yield from Vertosols. *Australian Journal of Agricultural Research, 59,* 321-330.

[5] Darwin, C. R. (1881). *The formation of vegetable mould through the action of worms, with observations on their habits.* London: John Murray.

[6] Derpsch, R. (2008). No-tillage and conservation agriculture: a progress report. In: T. Godard, M. A. Zoebisch, Y. T. Gan, W. Ellis, A. Watson & S. Sombatpanit (Eds.), *No-till Farming Systems Special Publication No. 3.* Bangkok: World Association of Soil and Water Conservation; 7-39.

[7] Douglas, L. A. (1987). Effects of cultivation and pesticide use on soil biology. In: P. S. Cornish & J. E. Pratley (Eds.), *Tillage - New Directions in Australian Agriculture.* Brisbane, Australia: Inkata; 308-317.

[8] Holt, J. A., Robertson, L. N. and Radford, B. J. (1993). Effects of tillage and stubble treatments on termite activity in two central Queensland Vertosols. *Australian Journal of Soil Research, 31,* 311-317.

[9] Kabir, Z. (2005). Tillage or no-tillage: impact on mycorrhizae. *Canadian Journal of Plant Science, 85,* 23-29.

[10] Li, H. W., Gao, H. W., Wu, H. D., Li, W. Y., Wang, X. Y. and He, J. (2007). Effects of 15 years of conservation tillage on soil structure and productivity of wheat cultivation in northern China. Australian Journal of Soil Research, *45,* 344-350.

[11] Radford, B. J. and Thornton, C. M. (2009). Effects and after-effects of 20 years of reduced tillage practices on soil properties and crop performance in the semi-arid subtropics. In: Nardali, E. T. (Ed.), *No-Till Farming: Effects on Soil, Pros and Cons and Potential* (??-??). New York, USA: Nova Science Publishers, Inc.

[12] Radford, B. J., Wilson-Rummenie, A. C., Simpson, G. B., Bell, K. L. and Ferguson, M. A. (2001). Compacted soil affects soil macrofauna populations in a semi-arid environment in central Queensland. *Soil Biology and Biochemistry, 33,* 1869-1872.

[13] Radford, B. J., Yule, D. F., McGarry, D. and Playford, C. (2007). Amelioration of soil compaction can take 5 years on a Vertisol under no-till in the semi-arid subtropics. *Soil and Tillage Research, 97,* 249-255.

[14] Rayment, G. E. and Bruce, R. C. (1984). *Soil testing and some soil test interpretations used by the Queensland Department of Primary Industries.* QDPI Information series QI84029. Brisbane, Australia: Queensland Department of Primary Industries.

[15] Rayment, G. E. and Higginson, F. R. (1992). *Australian Laboratory Handbook of Soil and Water Chemical Methods.* Melbourne, Australia: Inkata Press.

[16] Robertson, L. N. and Simpson, G. (1988). Sampling and dispersion of *Pterohelaeus alternatus* Pascoe and *Gonocephalum macleayi* (Blackburn) (Coleoptera: Tenebrionidae) larvae in soil. *Queensland Journal of Agricultural and Animal Sciences, 45,* 189-193.

[17] Robson, A. D. and Taylor, A. C. (1987). In: Cornish, P. S. and Pratley, J. E. (Eds.), *Tillage - New Directions in Australian Agriculture* (284-307). Melbourne, Australia: Inkata Press.

[18] Sá, J. C. M. (2004). Adubãcao Fosfatadano Sistema de Plantio Direto. In: T. Yamada, Silvia and R. S. Abdalla (Eds.), *Sympósio sobre Fósforo na Agricultura Brasileira, Anais* (201-222). Piracicaba, São Paulo: POTAFÓS.

[19] Standley, J., Hunter, H. M., Thomas, G. A., Blight, G. W. and Webb, A. A. (1990). Tillage and crop residue management affect Vertisol properties and grain sorghum growth over seven years in the semi-arid sub-tropics. 2. Changes in soil properties. *Soil and Tillage Research, 17,* 367-388.

[20] Swift M. J., Heal, O. W. and Anderson, J. M. (1979). *Decomposition in terrestrial ecosystems.* Blackwell, Oxford.

[21] Thomas, G. A., Standley, J., Webb, A. A., Blight, G. W. and Hunter, H. M. (1990a). Tillage and crop residue management affect Vertisol properties and grain sorghum growth over seven years in the semi-arid sub-tropics. 1. Crop growth, water use and nutrient balance. *Soil and Tillage Research, 17,* 181-197.

[22] Thomas, G. A., Standley, J., Hunter, H. M., Blight, G. W. and Webb, A. A. (1990b). Tillage and crop residue management affect Vertisol properties and grain sorghum growth over seven years in the semi-arid sub-tropics. 3. Crop residue and soil water during fallow periods. *Soil and Tillage Research, 17,* 389-407.

[23] Triplett, G. B. Jr. and Dick, W. A. (2008). No-tillage crop production: a revolution in agriculture. *Agronomy Journal, 100,* S-153–S-165.

[24] Wilson-Rummenie, A. C., Radford, B. J., Robertson, L. N., Simpson, G. B. and Bell, K. L. (1999). Reduced tillage increases population density of soil macrofauna in a semiarid environment in central Queensland. *Environmental Entomology, 28,* 163-172.

[25] Wright, A. L., Hons, F. M., Lemon, R. G., McFarland, M. L. and Nichols, R. L. (2007). Stratification of nutrients in soil for different tillage regimes and rotations. *Soil and Tillage Research, 96,* 19-27

In: New Developments in Biodiversity ... ISBN: 978-1-61324-374-9
Editor: Thomas W. Pace

Chapter 5

TOTAL METALLOTHIONEIN QUANTIFICATION BY REVERSED-PHASE HIGH-PERFORMANCE LIQUID CHROMATOGRAPHY COUPLED TO FLUORESCENCE DETECTION AFTER MONOBROMOBIMANE DERIVATIZATION

José Alhama[1*], Antonio Romero-Ruiz[2], Jamel Jebali[3] and Juan López-Barea[1]

[1]Department of Biochemistry and Molecular Biology, University of Córdoba, Severo Ochoa Building, 2nd floor, Campus de Rabanales, Highway A-4, Km 396a, 14071-Córdoba, Spain

[2]Instituto de Biomedicina de Sevilla, Hospital Universitario Virgen del Rocío, CSIC, Universidad de Sevilla, Avenida Manuel Siurot s/n, 41013 Sevilla, Spain

[3]Laboratoire de Biochimie et de Toxicologie Environnementale, Institut Supérieur Agronomique de Chott-Mariem, 4042 Sousse, Tunisia.

[*] Corresponing author: Tel.: +34 957 218082; Fax: +34 957 218688; E-mail address: bb2alcaj@uco.es

ABSTRACT

Metallothioneins (MTs) are ubiquitous and inducible proteins characterized by low molecular mass (Mr 6-8 kDa), high Cys content (20-30%) but no aromatic or His residues, and strong affinity to binding toxic metals (Cd, Hg, Ag, Pb) in metal-thiolate clusters. Due to their induction by a variety of stimuli, MTs are considered suitable biomarkers in the medical and environmental fields. The protective role of MTs from Cd toxicity and lethality is well-established. Although MT assessment is a difficult task, the accurate measurement of MT is mandatory in order to assess its biomarker potential and to identify new outstanding biological roles. We have developed a highly specific, sensitive, and reliable method for total MT quantification in unheated extracts by reversed-phase high-performance liquid chromatography coupled to fluorescence detection (RP-HPLC-FD). A derivatization protocol with monobromobinane, a thiol-specific fluorogenic reagent, is required after heat-, SDS- EDTA- and DTT-treatment. SDS-polyacrylamide gel electrophoresis was used to confirm the identity of the mBBr-labeled MT peak resolved by RP-HPLC-FD. The method has been successfully used to quantify MT content in the digestive gland of various clam species from Southern Spanish sites with different metal levels, and also in the liver of fish injected with different Cd, Cu and Hg doses. MT levels obtained by RP-HPLC-FD in non-heated extracts were significantly higher when compared to those obtained by other well-established assays relying on solvent precipitation (spectrophotometry) or heating (differential pulse polarography) pre-purification steps.

Keywords: biological roles; cadmium; exposure biomarker; fluorescence detection; reversed-phase high performance liquid chromatography; SDS-PAGE.

1. INTRODUCTION

Metallothioneins (MTs) were discovered in 1957 by Margoshes and Vallee while searching for a component responsible for the natural accumulation of Cd in mammalian kidney [1]. MTs were named according to their high metal content and unusually high number of Cys (20-30%) [2]. Also characterized by low-Mr (6-8 kDa) and absence of aromatic or His residues (Met in molluscs), MTs are inducible and ubiquitous proteins found in bacteria, fungi, plants and animal species [3, 4].

Experimental evidence suggests multiple biological functions for MTs, including: i) homeostasis of essential metals (Zn, Cu), ii) detoxification of essential and non-essential metals (Cd, Hg, Pt, Ag), and iii) antioxidant defense, by both free-radical scavenging and metal binding/release dynamics [2, 5-10]. In mammals, their involvement in metalloregulatory processes, including cell growth, differentiation and multiplication, has related MTs to carcinogenesis, from tumor cell pathology and drug resistance to apoptosis [9, 11]. MT expression also varies broadly in most pathological disorders in which metal metabolism is deregulated and reactive oxygen species are produced, including neurodegenerative diseases and senescence [7, 10, 12, 13]. Hence, MTs are considered to be suitable biomarkers in medicine, both of early diagnosis and of disease phase [7, 8, 11].

Nevertheless, several functions attributed to MTs are still subject to debate and the only role unequivocally established is protection from Cd toxicity [9, 10, 14, 15].

2. Cadmium and MTs

Exposure to toxic metals has become an increasingly recognized source of illness worldwide [16]. Cd is a ubiquitous pollutant of a great ecological and human concern [10, 17]. It is dramatically increasing due to its industrial uses, re-chargeable Ni-Cd batteries, zinc smelters, electroplating, pigment plants, plastic stabilizers, alloys, phosphate fertilizers [18]. Cigarette smoking is a major source of Cd exposure for the smoker, followed by diet for the non-smoker [18-20]. The Agency for Toxic Substances and Disease Registry lists Cd as being among the most hazardous substances in the environment [16], due to its wide range of organ toxicity and 10-30 years half-life [16, 21]. Adverse health effects of Cd exposure may occur at lower exposure levels than previously anticipated [18, 20]. Cd is easily absorbed and accumulated in important organs, mainly the kidneys and the liver [22-26]. This metal produces toxicity by various mechanisms: i) Alteration of sulfhydryl homeostasis by glutathione depletion and binding to protein –SH groups, which decreases antioxidant capacity by inhibiting antioxidant enzymes, alters the biological activities of many proteins, and disrupts the metabolism [16, 27, 28]. ii) Displacing Zn and Se in metalloenzymes thus decreasing their activity [16]. iii) Generating free radicals and lipid peroxidation [16, 29].

The mechanisms by which MT may protect cells against metal toxicity include: decreased uptake, metal sequestration, and enhanced export [9, 30].

Cd exposure has been associated with cancer and causes toxic effects in lung, kidney, liver, bone and immune system [18, 19, 23, 26]. Cd also induces apoptosis in many cell types through several mechanisms, including mitochondrial instability and oxidative stress [11, 29, 31]. Alterations are also apparent at the biochemical (carbohydrate and protein metabolism) and physiological levels [17, 22, 24-26, 32].

The protective role of MTs against Cd toxicity and lethality is well-established [6, 10, 14, 26, 28, 30, 33]. MTs protect cells from apoptosis induced by oxidative stress and metals (Cd) [9, 11, 34-36]. Cadmium is a particularly potent inducer of MT synthesis [33, 37]. In response to Cd, MTs are primarily induced and stored in the liver forming a complex, thus decreasing the Cd available to exert its toxic effect [16, 38-40]. Decreased uptake and enhanced metal export out of cells are other mechanisms by which MTs may protect from metal toxicity [9, 30]. MT levels showed the order liver>kidneys>gills, while MTmRNA had similar levels in the three tissues, implying differences in post-translational processes [40]. Exposure to low Zn doses was used in animal studies to induce MTs and protect from acute Cd-induced hepatotoxicity [16]. A CHO-K1 cell line continuously over-expressing MT was 13-fold more resistant to Cd effects than wild-type cells [14]. Mice genetically unable to produce MT are much more susceptible to renal damage and long-term Cd hepatotoxicity than MT-producing mice [16, 30]. It has been suggested that MT capacity to protect from Cd toxicity might have taken on a crucial role in the maintenance of human health and life processes, as compared to its other proposed functions [10, 30].

The induction of MTs synthesis by metals (Cd, Ag, Cu, Hg) has led to their proposed use as specific biomarkers for metal exposure and toxicity in aquatic biomonitoring [6, 7, 41-44]. MT was significantly induced by Cd (0-0.05 mg/L) in gills and digestive gland of the marine crab, *Chraybdis japonica*, with a dose-response relationship after 3 days′ exposure and a time-response relationship in digestive gland throughout the experimental period (15 days) [45]. Cd was a specially good inducer, increasing MT levels 6-fold over the control animals, in the shore crab *Carcinus maenas* [46]. Mussels, *M. galloprovincialis* exposed to Cd (1mg/L) for 1 week showed a 5-fold (MLP2) or 7-fold (MLP1) increase compared to non-exposed animals [47]. In *Sparus aurata* injected with 500 μg/kg of Cd for 2 days, MT levels increased significantly in liver, gills and kidney, with the highest value (3.3-fold) in liver [39]. Many other reports have shown that MTs are induced in Cd-exposed fish, such as hybrid tilapia *Oreochromis* sp. [24], common carp *Cyprinus carpio* [22], rainbow trout *Oncorhynchus mykiss* [38], sea bass *Dicentrarchus labrax*

[25] and the turbot *Scophthalmus maximus* [40]. The role of MTs in the detoxification of harmful metals permits the acquisition of metal tolerance for organisms living in metal-contaminated environments [42, 48].

3. Quantification of MTs by RP-HPLC with Fluorescence Detection

It has long been assumed that MTs might play a key physiological function. Yet, despite enormous and multidisciplinary efforts involving structural biochemistry and molecular biology studies, the primary role of MTs remains elusive, in spite of further functions being incessantly assigned [15, 36, 49]. For this reason, the precise measurement of MT is mandatory in order to assess its biomarker potential and to identify new outstanding biological roles [7, 50].

Even though many techniques and methodologies have been developed for purification and quantification of total MTs, reported concentrations differ widely between laboratories and are expressed in different units [6, 44, 50]. A large number of different, non-intercalibrated protocols have been used to measure MTs in various organisms, making it difficult to compare results [44, 51, 52]. It is important to highlight that to compare MT levels to data from the literature it is necessary to take into consideration the isolation and quantification procedure [52].

Investigation and quantification of MTs is limited by their absence of biological activity and of the typical absorption and emission spectra of proteins, due to their lack of aromatic amino acids. Nevertheless, their immunological properties [50] as well as their high metal-binding capacity and elevated Cys content have been used as distinctive features for the development of different analysis methods [44, 53]. Additionally, the levels of MT isoforms have been evaluated in the Mytilus genus using real-time quatitative-PCR methods [54], based on its available DNA sequence, although the use of this Molecular Biology approach is more challenging in non-model species with unknown MT sequences. Methods based on their abundant thiol groups are chiefly used in marine bivalves, including Hg-saturation, differential pulse polarography (DPP) or spectrophotometry [44]. Sensitivity is enhanced by reaction of thiol groups with a fluorogenic compound [55, 56]. Protein thiols can be labeled with several reagents, including DBPM (N-[4-dimethylamino-2-benzofuranyl) phenyl]maleimide) [57], SBD-F (ammonium

7-fluorobenz-2-oxa-1,3-diazole-4-sulfonate) and ABD-F (4-aminosulfonyl-7-fluoro-2,1,3-benzoxadiazole) [58], 5-idoacetamide-fluoescein [59] and mBBr (monobromobimane) [60]. Several groups have reported that fluorescent detection, after the reaction of a fluorogenic reagent with MT thiol groups, combined with chromatographic separation, yields a high resolution and sensitivity. Thus, gel permeation and DBPM labeling allowed to assay MTs in rat tissues [57]. Isocratic HPLC was used for MT quantification of diverse organisms after SBD-F coupling; two different reverse-phase columns in tandem resolved SBD-labeled MT from other SBD complexes [61]. An HPLC-fluorescence method was adapted for MT quantification in *Mytilus galloprovincialis* after a two-step acetone precipitation using SBD-F and an acetonitrile gradient [47].

We have developed an easy, highly sensitive and specific method to assess total MT content by RP-HPLC coupled to fluorescent detection after mBBr labeling [55, 56]. Fluorescence of mBBr-labeled proteins was measured with excitation at 382 nm and emission at 470 nm, using rabbit liver MT-I as a reference standard. The reactivity of mBBr is well-established and its allylic bromide reacts fast and specifically with thiol groups to yield highly fluorescent derivatives [62, 63], while the parental reagent is essentially non-fluorescent [60]. Prior to RP-HPLC-FD, a derivatization step with mBBr of metal-depleted MT thiols is required after EDTA, SDS and DTT treatment at 70ºC. Optimal concentrations of DTT (2 mM), SDS (3%) and mBBr (12 mM) were established when labeling conditions were studied using rabbit MT-I as standard; labeling conditions were the same for clam extracts except that a higher DTT (~12 mM) concentration was required, suggesting the presence in the extracts of high oxidant levels (Table 1) [56]. Importance of SDS purity in MT derivatization has also been highlighted. Only with high-purity SDS (>99%) mBBr-MT from a digestive gland clam extract eluted in a sharp well-defined peak, while at least three peaks were obtained using 95% SDS [56]. Since the derivatization of thiol groups with mBBr occurs mainly in the dissociated thiolate form [60], the effect of pH on mBBr derivatization of MTs was studied. Maximum fluorescent labeling was obtained at pH 8.5-9.5 [55]. The mBBr-MT complex was noticeably stable even after 25 h incubation at room temperature [55], since, unlike other labeling reagents, mBBr fluorescence does not fade appreciably with time [62, 64].

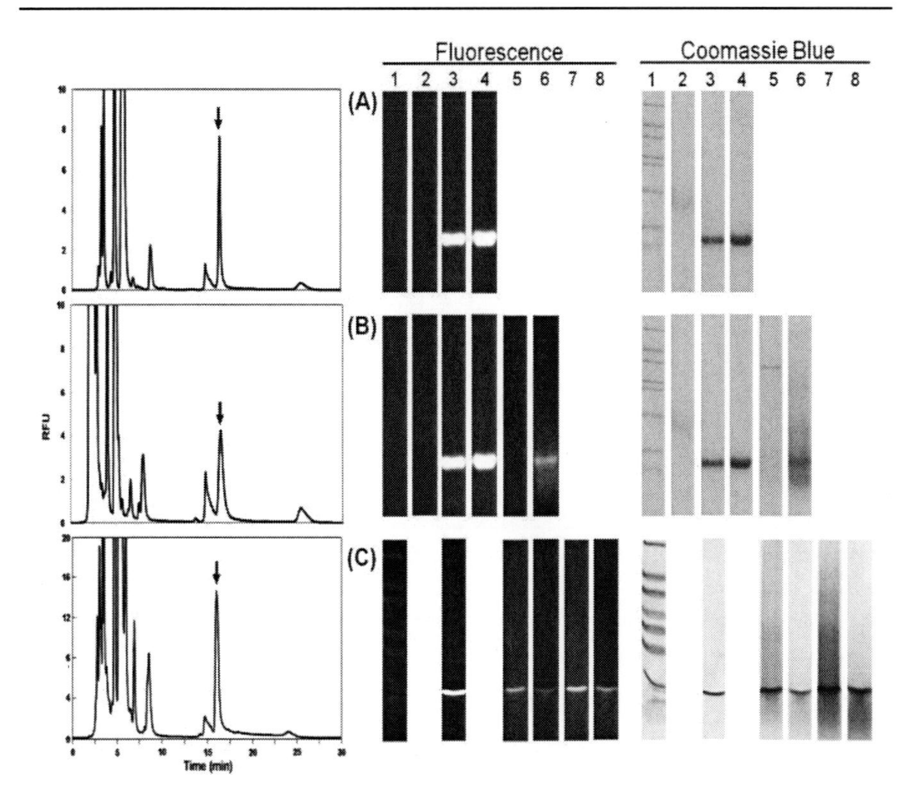

Figure 1. (Left) RP-HPLC-FD elution profiles of mBBr derivatives of purified rabbit liver MT-I (A), heated digestive gland extract of *C. gallina* (B), and non-heated digestive gland extract of *S. plana* (C). Arrows show MT peaks. (Right) Analysis of rabbit liver MT-I (A) and of MTs from *C. gallina* extract (B) in 15% Tris-Glycine SDS-PAGE gels, and from *S. plana* extract (C) in 13.5% Tris-Tricine SDS-PAGE gels. Fluorescence of mBBr-labeled and absorbance of Coomassie-stained proteins were assessed in the same gel. Lane 1, 14 μg of unlabeled (A and B) and 30 μg of mBBr-labeled (C) standard proteins; lane 2, 5 μg of unlabeled rabbit MT-I; lane 3, 5 μg (A and B) and 1.3 μg (C) of mBBr-MT-I from rabbit; lane 4, 10 μg of rabbit mBBr-MT-I after RP-HPLC; lane 5, 52 μg (*C. gallina*) and 30 μg (*S. plana*) of total mBBr-labeled protein from heated clam extract; lane 6, 12 μg (*C. gallina*) and 7 μg (*S. plana*) of mBBr-labeled MT from heated clam extract after RP-HPLC; lane 7, 30 μg of total mBBr-labeled protein from non-heated *S.plana* clam extract; lane 8, 12 μg of mBBr-labeled MT purified by RP-HPLC from non-heated *S.plana* extract. Chromatographic and electrophoretic conditions are described in Alhama et al [55] and Romero-Ruiz et al [56]. [Modified from Figures 3 and 6 in Alhama et al (2006) *J Chromatogr A. 1107*, 52-8 and from Figures 3 and 4 in Romero-Ruiz et al (2008) *Environ Pollut. 156*, 1340-7].

Table 1. Optimization of MT labeling with mBBr[a]

DTT (mM)	MT-I (purified rabbit liver)	Clam (digestive gland)
0	98.3	4.1
0.5	82.8	-
1	100.0	-
2	96.4	6.6
3	87.2	-
4	67.9	18.6
6	39.3	41.9
8	-	79.7
10	-	100.0
12	-	98.6
SDS (%)	**MT-I (purified rabbit liver)**	**Clam (digestive gland)**
0	0.0	0.0
0.5	43.1	48.6
1	69.4	72.8
2	80.5	89.2
3	100.0	100.0
4	83.1	90.9
6	78.5	76.3
mBBr (mM)	**MT-I (purified rabbit liver)**	**Clam (digestive gland)**
6	37.9	97.2
9	86.2	91.0
12	100.0	100.0
15	84.7	63.0
18	81.4	83.4
21	70.0	89.7

[a] Purified rabbit liver MT and *Scrobicularia plana* clam digestive gland extracts were heated at 70ºC, derivatized with mBBr and analyzed by RP-HPLC-FD. Values are shown as percentage of maximum. [Modified from Figure 2 in Romero-Ruiz et al (2008) *Environ Pollut. 156*, 1340-7].

Table 2. Metallothionein content and metal levels in *Scrobicularia plana* clams from different Guadalquivir Estuary sites

Sampling site [b]	MT content [a] (mg/g protein)		Metal content (μg/g wet weight)								
	Heated	Non-heated	Zn	Pb	Cd	Ni	Mn	Fe	Cu	As	Cr
BT	62.6 ± 2.0	139.6 ± 5.7	41.9 ± 2.6	2.18 ± 0.11	0.126 ± 0.010	0.85 ± 0.05	15.6 ± 0.3	1973 ± 95	5.65 ± 0.46	2.13 ± 0.14	0.50 ± 0.03
SRs	80.4 ± 9.8	199.4 ± 12.3 **	66.7 ± 1.2 **	2.56 ± 0.18	0.096 ± 0.008 *	1.33 ± 0.10 **	20.9 ± 0.8 **	2833 ± 234 **	7.92 ± 0.41 *	2.99 ± 0.20 **	1.03 ± 0.09 **
Bh	84.3 ± 10.4 * (c)	192.1 ± 12.0 **	107.0 ± 1.0 **	4.02 ± 0.24 **	0.109 ± 0.009	1.79 ± 0.09 **	27.2 ± 2.3 **	3173 ± 278 **	14.00 ± 1.10 **	3.91 ± 0.15 **	1.16 ± 0.03 **

[a] MT analysis was carried out by RP-HPLC-FD in heated and non-heated digestive gland extracts.

[b] Clams were sampled on October 2003 at three sites: “Brazo de la Torre” (BT), “San Rafael” salt works (SRs), and across Bonanza harbor (Bh).

[c] Statistical significances for comparison with BT are indicated as follows: *, $p < 0.05$; **, $p < 0.01$.

[Modified from Figure 5 and Table 2 in Romero-Ruiz et al (2008) *Environ Pollut. 156*, 1340-7].

After optimizing the chromatographic conditions, mBBr-labeled MT eluted in a well-defined fluorescent peak using a linear 30-70% acetonitrile gradient (in the presence of 0,1% TFA) well separated from unreacted mBBr, mBBr-labeled DTT, and other thiol-containing proteins, that all eluted at the initial conditions (30% acetonitrile) in the void volume of the HPLC column (Figure 1, left) [55]. Although MTs have long been considered to be heat-resistant proteins [2, 44, 65], we compared MT levels in digestive gland extracts heated at 95ºC for 10 min and in non-heated extracts. While MT patterns were similar, its content was much higher in unheated extracts, suggesting that heating removes part of the MT initially present in the extracts [56]. These results were confirmed by electrophoretic analysis (Figure 1, right). In heated extracts after RP-HPLC, a sharp and highly fluorescent band was visible (lane 6) with Mr slightly higher than rabbit liver MT-I, while a more intense band appeared in unheated purified extracts (lane 8). The same differences were observed when extracts not purified by RP-HPLC were compared (lanes 5 and 7). It has been reported that MT co-precipitates during heat-treatment, leading to its underestimation [51, 66]. The RP-HPLC-FD assay allows a direct assessment of total MT in extracts without heat treatment, since mBBr-labeled MT is well separated by RP-HPLC from other Cys containing proteins, as shown by electrophoretic fluorimetric analysis, obtaining a single intense fluorescent band after chromatography of non-heated clam extracts (Figure 1, right; lane 8) [56].

The new RP-HPLC-FD method has been successfully used to evaluate total MT content in clam digestive gland from sites with different metal levels (Table 2) [55, 56] and in sea bass specimens injected with different Cd, Cu and Hg doses (Figure 2) [67]. *Chamaelea gallina* clams from different Huelva coastal sites had significantly higher MT levels than those from the reference site, according to the higher metal contents of Huelva animals [55, 68]. MT and other well-established biomarkers were measured in *Scrobicularia plana* clams to assess pollution of the Guadalquivir Estuary (SW Spain). Significantly, MT content determined by RP-HPLC-FD in non-heated extracts was much higher (30-148% increase) than that obtained in heated extracts [56]. Table 2 shows the metal levels and MT content in heated and non-heated digestive gland extracts of clams sampled on October 2003 at three sites of the Guadalquivir Estuary. MT content assayed by RP-HPLC-FD in non-heated samples was well-correlated with metals (significant positive correlation with Zn, Pb, Ni, Mn and Fe) and anti-oxidant activities (significant negative correlation with 6PGDH and Glyox II, and significant positive correlation with catalase and GST). In contrast, none of the metals or biomarkers showed a significant correlation with MT levels in heated extracts [56]. We would like to emphasize the extremely high MT levels found in *S. plana*, 89-199 mg g^{-1} protein, representing 9-20% of total soluble proteins. The key cell functions of MTs, especially in bivalve mollusks living in environments with high metal levels, would explain the high concentration of this protein [56]. In exposed *Dicentrarchus labrax*, liver MT increased linearly with Cu and Hg doses, and was saturated beyond 100 µg kg^{-1} Cd. Maximum induction was obtained at 100 µg kg^{-1} Cd (5.3-fold), and 250 µg kg^{-1} Cu or Hg (8- and 5.1-fold, respectively) (Figure 2) [67]. MT contents do not always reflect the Cd levels in fish. At the higher Cd dose studied (250 µg kg^{-1}), synthesis of hepatic MT was clearly reduced, becoming limiting due to a progressive inhibition of critical metabolic processes (citotoxicity) [23, 24, 33, 40, 48, 67, 69], according to other reported studies in the turbot (*Scothpthalmus maximus*) [40] and the greater amberjack (*Seriola dumerilli*) [69]. Compared to the spectrophotometric assay that titrates the –SH groups released from metal-striped MT with Ellman´s reagent [70], the RP-HPLC-FD method detected significantly higher total MT content (1.31-1.95-fold) in all metal-exposed animals [67]. The discrepancy between both methods could be attributed to an underestimation of the MT content due to the use of GSH as standard instead of MTs, and/or to partial co-precipitation of MT with hydrophobic proteins during the solvent extraction required before the spectrophotometric assay [44, 51, 55, 67].

It has been shown that heat treatment and solvent precipitation effectively remove high molecular weight proteins that interfere with MT determination by several methods [51]. However, MT isoforms present different thermal stability and resistance to oxidation and polymerization by organic solvents. Thus, the mussel *Mytilus galloprovincialis* displays two MT isoforms: MT-10 is a housekeeping protein preferentially induced by essential metals, whereas MT-20 is highly induced by Cd. This last isoform has a less compact structure and it is significantly depleted by heat treatment and drastically diminished by solvent precipitation [49, 51]. Important MT losses were also clearly established in heated clam extracts [56]. The MT gene expression levels, assayed by quantitative real-time PCR in fish and mollusks contaminated with Cd, increased further beyond MT protein levels, with the metal accumulated at tissue level clearly overtaking the sequestration capacities of MT [54, 71]. Dondero et al [54] suggested that posttranscriptional control mechanisms could explain the discrepancy between MT transcriptional induction and protein accumulation. However, underestimation of MT during purification and assessment at the protein level should not be ruled out. In consequence, methods that require a pre-purification protocol based on heat treatment or solvent precipitation could underestimate MT levels, thus making it difficult to reveal the biological roles of this elusive protein. In contrast, in the RP-HPLC-FD method MTs are separated by HPLC as shown by electrophoretic fluorimetric analysis, obtaining a unique intense fluorescent band after chromatography of non-heated clam extracts. Thus, the MT levels obtained by RP-HPLC-FD in non-heated extracts were significantly higher when compared to those obtained by other well-established methods, the DPP electrochemical method [56] and the spectrophotometric assay [67], that rely on heating and solvent precipitation pre-purification steps, respectively.

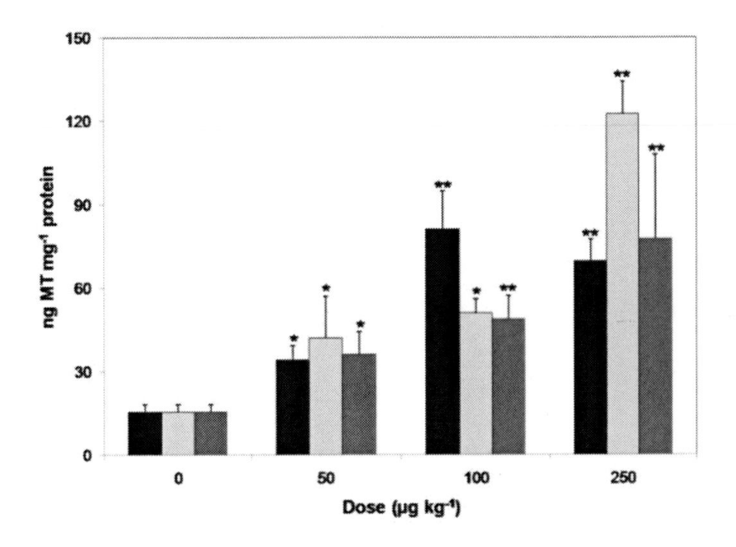

Figure 2. Metallothionein content determined by RP-HPLC-FD in the liver of the sea bass *D. labrax* injected with different Cd (■), Cu (■) and Hg (■) concentrations after 48 h exposure. Statistical significance of the differences between exposed and control groups are shown as; *, $p<0.05$; **, $p<0.01$. [Modified from Figure 3 in Jebali et al (2008) *Mar Environ Res. 65*, 358-63].

4. Electrophoretic Analysis of MTs

SDS-PAGE studies have not usually been performed in MTs since these proteins behave anomalously in this electrophoresis system [72-75]. MT patterns are sometimes obtained as broad, faint bands of low mobility, and with a limited binding to blotting membranes [55, 63, 76]. Like other Cys-rich proteins, MTs aggregate, mainly in dimers, by forming disulfide bonds due to thiol oxidation during the stacking phase of SDS-PAGE [55, 72, 76-78].

A number of *in vitro* studies have shown that MTs dimerize through disulfide bond formation under oxidative conditions [9, 79]. These dimers have also been described to occur *in vivo* under conditions of oxidative stress or when animals were exposed to Cd [7, 52, 80]. Oxidation and degree of polymerization have been analyzed by labeling MT thiols with eosin-5-iodoacetamide followed by SDS-PAGE. Higher-order MT aggregates via intermolecular disulfides might be physiologically important for subcellular retention, protection from degradation, and/or storage [75]. Cysteine oxidation was also proposed to be involved in the dissociative mechanism controlling free Zn fluctuations and modulation of cell signaling pathways [81, 82]. A chimeric MT, mimicking the natural dimeric form, was recently constructed to show that *in vivo*-formed dimers have a greater Cd sequestering capacity [80]. In fact, two major groups of MTs have been identified in Mytilidae, a monomeric constitutive isoform (MT-10) and a dimer (MT-20) which is mainly induced by Cd [49, 51, 54, 83].

The aggregation can be prevented by reduction of thiol groups followed by irreversible blockage with iodoacetamide [76, 77], SBD-F [47], eosin-5-iodoacetamide [75] or mBBr [55, 56, 63, 70]. Carboxymethylated MTs can be detected after SDS-PAGE by using silver staining and autoradiography [76]. SDS-PAGE has also been used to identify MTs labeled with SBD-F and purified by a two-step acetone precipitation [47]. Monobromobimane (mBBr) is a fluorescent probe for both protein and non-protein thiols [60, 62]. Fluorescent identification of mBBr-labeled MTs after SDS-PAGE was reported after purification and concentration by acidic ethanol-chloroform of mussel digestive gland [70]. Using the same approach, in addition to a constitutive MT band, a significant increase in a MT peptide was observed only after Cd (not Cu, Hg or Zn) exposure in the mussel *M. galloprovincialis* [54]. SDS-PAGE was used to confirm the identity of the mBBr-labeled MT peak resolved by RP-HPLC-FD (Figure 1, right) [55, 56]. The unique properties of MTs, low-Mr and high Cys content, permitted their correct identification. Due to its small size, improved separation and resolution of MTs in SDS-PAGE is obtained using the Tricine discontinuous buffer system described by Schagger and von Jagow (Figure 1C, right) [55, 70, 84]. Mr standards can also be labeled with mBBr; taking advantage of the introduced fluorescence groups, the modified proteins can also be visualized directly in the gel by fluorescence imaging (Figure 1C, right; lane 1) [56].

Labeling with mBBr has multiple advantages for visualizing MTs in gel electrophoresis: i) mBBr labeling of MTs block their Cys-SH, leading to formation of a sharp monomer band [55, 56, 63, 70]. ii) Due to its high Cys content and the hydrophobicity of the mBBr introduced, the properties of the labeled protein are notably altered [55, 63]. Since MTs lack aromatic residues, they are not efficiently stained by Coomassie blue, which selectively binds through Van der Waals forces and hydrophobic contacts with the aromatic amino acids Trp, Tyr, and Phe [60]. After mBBr labeling, the higher hydrophobicity of mBBr-MT explains its increased Coomassie staining (Figure 1, right) [55]. iii) The sensitivity of fluorescent detection often surpasses that of conventional staining procedures [62]. iv) MTs can be digitally imaged directly post-electrophoresis, thus obviating lengthy and expensive staining procedures [75, 85]. Since the proteins are pre-labeled, this reagent allows their visualization during electrophoresis. Additionally, gels do not have to be manipulated after electrophoresis and, like in Difference Gel Electrophoresis (DIGE), they could be imaged within the glass plates [85]. This could be advantageous for 2-DE, reducing the variation between gels, and the risk of damaging or destroying them, mainly when working with large format gels that are cumbersome to handle. v) Modification of MTs with mBBr improves blotting efficiency, resulting in a highly sensitive detection in SDS-PAGE and Western blots [63].

Two-dimensional polyacrylamide gel electrophoresis (2-DE) analysis is proposed for higher resolution of MTs. Nevertheless, the anomalous behavior of MTs limits their proteomic analysis; thus, although their thiol groups are usually reduced and later blocked with iodoacetamide for the second dimension (SDS-PAGE), the first dimension (isoelectrofocusing, IEF) is carried out in their native form; further work is needed to clarify MTs behavior in the IEF gel. It should be noted that MTs are not usually identified after 2-DE. We only know one paper describing the identification of an MT-like protein after 2-DE of rice seeds germinated with a toxic Cu concentration [86]. MS-based analysis is a popular and powerful tool for protein identification and characterization after 2-DE. However, metal-saturated MTs are resistant to proteolytic digestion by several enzymes including trypsin, the usual proteolytic agent in proteomics, making its proteomic identification difficult. By adding EDTA to the samples, the problem is overcome by rendering MTs readily digested into peptides and identified by MS/MS [87]. A method was described whereby proteins containing thiol groups are labeled with mBBr prior to IEF of 2-DE; high resolution spot patterns were imaged while the gels were still in their glass cassettes. A high spot count (approx. 10%) on the fluorescent gels indicated the detection of spots undetected by silver staining [85]. 2-DE has also been used to separate mBBr-labeled proteins after *in vitro* reduction by NADP/thioredoxin to identify thioredoxin targets in developing seeds [88, 89]. Our group is currently involved in several projects aimed at evaluating metal pollution at Doñana National Park (SW Spain) and along the Tunisian coast, using mice, crabs, earthworms and soil microorganisms as bioindicators. 2-DE analysis of MTs after mBBr labeling has been proposed for these projects. Besides, because proteins are separated by IEF (pI) in the first dimension, different MT isoforms could hopefully be well-resolved. MT analysis by 2-DE could help us to prove some of the debatable functions of MTs, and to elucidate and assign new roles to this elusive protein.

ACKNOWLEDGMENT

This work was funded by grant CMT2006-08960-C02 from the Spanish Ministry of Education and Science, a grant (P08-CVI-03929) from the Agency of Innovation, Science and Enterprise (Andalusian Regional Government) and by a mobility grant (A/016113/08) from the Spanish Ministry of Foreign Affairs, Spanish Agency of International Cooperation, in the Spain-Tunisia program.

REFERENCES

[1] Margoshes, M. and Vallee, B. L. (1957) A cadmium protein from equine kidney cortex, *J Am Chem Soc. 79*, 1813-14.

[2] Kagi, J. H. (1991) Overview of metallothionein, *Methods Enzymol. 205*, 613-26.

[3] Kojima, Y. (1991) Definitions and nomenclature of metallothioneins, *Methods Enzymol. 205*, 8-10.

[4] Kojima, Y., Binz, P. A. and Kägi, J. H. R. (1999) Nomenclature of metallothionein: Proposal for a revision in *Metallothionein IV* (Klaassen, C., ed) pp. 3-6, Birkhauser Verlag, Basel.

[5] Viarengo, A., Burlando, B., Ceratto, N. and Panfoli, I. (2000) Antioxidant role of metallothioneins: a comparative overview, *Cell Mol Biol (Noisy-le-grand). 46*, 407-17.

[6] Amiard, J. C., Amiard-Triquet, C., Barka, S., Pellerin, J. and Rainbow, P. S. (2006) Metallothioneins in aquatic invertebrates: their role in metal detoxification and their use as biomarkers, *Aquat Toxicol. 76*, 160-202.

[7] Carpene, E., Andreani, G. and Isani, G. (2007) Metallothionein functions and structural characteristics, *J Trace Elem Med Biol. 21 Suppl 1*, 35-9.

[8] Vasak, M. (2005) Advances in metallothionein structure and functions, *J Trace Elem Med Biol. 19*, 13-7.

[9] Formigari, A., Irato, P. and Santon, A. (2007) Zinc, antioxidant systems and metallothionein in metal mediated-apoptosis: biochemical and cytochemical aspects, *Comp Biochem Physiol C Toxicol Pharmacol. 146*, 443-59.

[10] Klaassen, C. D., Liu, J. and Choudhuri, S. (1999) Metallothionein: an intracellular protein to protect against cadmium toxicity, *Annu Rev Pharmacol Toxicol. 39*, 267-94.

[11] Thirumoorthy, N., Manisenthil Kumar, K. T., Shyam Sundar, A., Panayappan, L. and Chatterjee, M. (2007) Metallothionein: an overview, *World J Gastroenterol. 13*, 993-6.

[12] Maret, W. (2008) A role for metallothionein in the pathogenesis of diabetes and its cardiovascular complications, *Mol Genet Metab. 94*, 1-3.

[13] Maret, W. (2008) Metallothionein redox biology in the cytoprotective and cytotoxic functions of zinc, *Exp Gerontol. 43*, 363-9.

[14] Beattie, J. H., Owen, H. L., Wallace, S. M., Arthur, J. R., Kwun, I. S., Hawksworth, G. M. and Wallace, H. M. (2005) Metallothionein overexpression and resistance to toxic stress, *Toxicol Lett. 157*, 69-78.

[15] Coyle, P., Philcox, J. C., Carey, L. C. and Rofe, A. M. (2002) Metallothionein: the multipurpose protein, *Cell Mol Life Sci. 59*, 627-47.

[16] Patrick, L. (2003) Toxic metals and antioxidants: Part II. The role of antioxidants in arsenic and cadmium toxicity, *Altern Med Rev. 8*, 106-28.

[17] Giari, L., Manera, M., Simoni, E. and Dezfuli, B. S. (2007) Cellular alterations in different organs of European sea bass *Dicentrarchus labrax* (L.) exposed to cadmium, *Chemosphere. 67*, 1171-81.

[18] Jarup, L. (2002) Cadmium overload and toxicity, *Nephrol Dial Transplant. 17 Suppl 2*, 35-9.

[19] Jarup, L., Berglund, M., Elinder, C. G., Nordberg, G. and Vahter, M. (1998) Health effects of cadmium exposure-a review of the literature and a risk estimate, *Scand J Work Environ Health. 24 Suppl 1*, 1-51.

[20] Jarup, L. (2003) Hazards of heavy metal contamination, *Br Med Bull. 68*, 167-82.

[21] Stoeppler, M. (1991) Cadmium in *Metals and Their Compounds in the Environment. Occurrence, Analysis and Biological Relevance* (Merian, E., ed) pp. 804-51, VCH, Weinheim.

[22] De Smet, H. and Blust, R. (2001) Stress responses and changes in protein metabolism in carp *Cyprinus carpio* during cadmium exposure, *Ecotoxicol Environ Saf. 48*, 255-62.

[23] Berntssen, M. H., Aspholm, O. O., Hylland, K., Wendelaar Bonga, S. E. and Lundebye, A. K. (2001) Tissue metallothionein, apoptosis and cell proliferation responses in Atlantic salmon (*Salmo salar* L.) parr fed elevated dietary cadmium, *Comp Biochem Physiol C Toxicol Pharmacol. 128*, 299-310.

[24] Wu, S. M., Shih, M. J. and Ho, Y. C. (2007) Toxicological stress response and cadmium distribution in hybrid tilapia (*Oreochromis* sp.) upon cadmium exposure, *Comp Biochem Physiol C Toxicol Pharmacol. 145*, 218-26.

[25] Cattani, O., Serra, R., Isani, G., Raggi, G., Cortesi, P. and Carpene, E. (1996) Correlation between metallothionein and energy metabolism in sea bass, *Dicentrarchus labrax*, exposed to cadmium, *Comp Biochem Physiol C. 113*, 193-9.

[26] Swiergosz-Kowalewska, R. (2001) Cadmium distribution and toxicity in tissues of small rodents, *Microsc Res Tech. 55*, 208-22.

[27] Valko, M., Morris, H. and Cronin, M. T. (2005) Metals, toxicity and oxidative stress, *Curr Med Chem. 12*, 1161-208.

[28] Quig, D. (1998) Cysteine metabolism and metal toxicity, *Altern Med Rev. 3*, 262-70.

[29] Pulido, M. D. and Parrish, A. R. (2003) Metal-induced apoptosis: mechanisms, *Mutat Res. 533*, 227-41.

[30] Park, J. D., Liu, Y. and Klaassen, C. D. (2001) Protective effect of metallothionein against the toxicity of cadmium and other metals, *Toxicology. 163*, 93-100.

[31] Hamada, T., Tanimoto, A. and Sasaguri, Y. (1997) Apoptosis induced by cadmium, *Apoptosis. 2*, 359-67.

[32] Romeo, M., Bennani, N., Gnassia-Barelli, M., Lafaurie, M. and Girard, J. P. (2000) Cadmium and copper display different responses towards

oxidative stress in the kidney of the sea bass *Dicentrarchus labrax*, *Aquat Toxicol. 48*, 185-94.

[33] Bremner, I. and Beattie, J. H. (1990) Metallothionein and the trace minerals, *Annu Rev Nutr. 10*, 63-83.

[34] Shimoda, R., Achanzar, W. E., Qu, W., Nagamine, T., Takagi, H., Mori, M. and Waalkes, M. P. (2003) Metallothionein is a potential negative regulator of apoptosis, *Toxicol Sci. 73*, 294-300.

[35] Shimoda, R., Nagamine, T., Takagi, H., Mori, M. and Waalkes, M. P. (2001) Induction of apoptosis in cells by cadmium: quantitative negative correlation between basal or induced metallothionein concentration and apoptotic rate, *Toxicol Sci. 64*, 208-15.

[36] Vasak, M. and Hasler, D. W. (2000) Metallothioneins: new functional and structural insights, *Curr Opin Chem Biol. 4*, 177-83.

[37] Webb, M. (1986) Role of metallothionein in cadmium metabolism, *Handb Exp Phamacol. 80*, 281-337.

[38] Castano, A., Carbonell, G., Carballo, M., Fernandez, C., Boleas, S. and Tarazona, J. V. (1998) Sublethal effects of repeated intraperitoneal cadmium injections on rainbow trout (*Oncorhynchus mykiss*), *Ecotoxicol Environ Saf. 41*, 29-35.

[39] Ghedira, J., Jebali, J., Bouraoui, Z., Banni, M., Guerbej, H. and Boussetta, H. (2010) Metallothionein and metal levels in liver, gills and kidney of *Sparus aurata* exposed to sublethal doses of cadmium and copper, *Fish Physiol Biochem. 36*, 101-7.

[40] George, S. G., Todd, K. and Wright, J. (1996) Regulation of metallothionein in teleosts: induction of MTmRNA and protein by cadmium in hepatic and extrahepatic tissues of a marine flatfish, the turbot (*Scophthalmus maximus*), *Comp Biochem Physiol C Pharmacol Toxicol Endocrinol. 113*, 109-15.

[41] Cajaraville, M. P., Bebianno, M. J., Blasco, J., Porte, C., Sarasquete, C. and Viarengo, A. (2000) The use of biomarkers to assess the impact of pollution in coastal environments of the Iberian Peninsula: a practical approach, *Sci Total Environ. 247*, 295-311.

[42] Monserrat, J. M., Martinez, P. E., Geracitano, L. A., Amado, L. L., Martins, C. M., Pinho, G. L., Chaves, I. S., Ferreira-Cravo, M., Ventura-Lima, J. and Bianchini, A. (2007) Pollution biomarkers in estuarine animals: critical review and new perspectives, *Comp Biochem Physiol C Toxicol Pharmacol. 146*, 221-34.

[43] Langston, W. J., Bebiano, M. J. and Burt, G. R. (1998) Metal binding strategies in molluscs in *Metal Metabolism in Aquatic Environments*

(Langston, W. J. and Bebiano, M. J., eds) pp. 219-83, Chapman and Hall, London.

[44] Cosson, R. P. (2000) Bivalve metallothionein as a biomarker of aquatic ecosystem pollution by trace metals: limits and perspectives, *Cell Mol Biol (Noisy-le-grand). 46*, 295-309.

[45] Pan, L. and Zhang, H. (2006) Metallothionein, antioxidant enzymes and DNA strand breaks as biomarkers of Cd exposure in a marine crab, *Charybdis japonica*, *Comp Biochem Physiol C Toxicol Pharmacol. 144*, 67-75.

[46] Pedersen, S. N., Pedersen, K. L., Hojrup, P., Knudsen, J. and Depledge, M. H. (1998) Induction and identification of cadmium-, zinc- and copper-metallothioneins in the shore crab *Carcinus maenas* (L.), *Comp Biochem Physiol C Pharmacol Toxicol Endocrinol. 120*, 251-9.

[47] El Ghazi, I., Menge, S., Miersch, J., Chafik, A., Benhra, A., Elamrani, M. K. and Krauss, G. J. (2003) Quantification of metallothionein-like proteins in the mussel *Mytilus galloprovincialis* using RP-HPLC fluorescence detection, *Environ Sci Technol. 37*, 5739-44.

[48] Roesijadi, G. (1992) Metallothioneins in metal regulation and toxicity in aquatic animals, *Aquat Toxicol. 22*, 81-114.

[49] Vergani, L., Grattarola, M., Grasselli, E., Dondero, F. and Viarengo, A. (2007) Molecular characterization and function analysis of MT-10 and MT-20 metallothionein isoforms from *Mytilus galloprovincialis*, *Arch Biochem Biophys. 465*, 247-53.

[50] Dabrio, M., Rodriguez, A. R., Bordin, G., Bebianno, M. J., De Ley, M., Sestakova, I., Vasak, M. and Nordberg, M. (2002) Recent developments in quantification methods for metallothionein, *J Inorg Biochem. 88*, 123-34.

[51] Erk, M., Ivankovic, D., Raspor, B. and Pavicic, J. (2002) Evaluation of different purification procedures for the electrochemical quantification of mussel metallothioneins, *Talanta. 57*, 1211-18.

[52] Isani, G., Andreani, G., Kindt, M. and Carpene, E. (2000) Metallothioneins (MTs) in marine molluscs, *Cell Mol Biol (Noisy-le-grand). 46*, 311-30.

[53] Haase, H. and Maret, W. (2004) A differential assay for the reduced and oxidized states of metallothionein and thionein, *Anal Biochem. 333*, 19-26.

[54] Dondero, F., Piacentini, L., Banni, M., Rebelo, M., Burlando, B. and Viarengo, A. (2005) Quantitative PCR analysis of two molluscan

metallothionein genes unveils differential expression and regulation, *Gene. 345*, 259-70.

[55] Alhama, J., Romero-Ruiz, A. and Lopez-Barea, J. (2006) Metallothionein quantification in clams by reversed-phase high-performance liquid chromatography coupled to fluorescence detection after monobromobimane derivatization, *J Chromatogr A. 1107*, 52-8.

[56] Romero-Ruiz, A., Alhama, J., Blasco, J., Gomez-Ariza, J. L. and Lopez-Barea, J. (2008) New metallothionein assay in *Scrobicularia plana*: heating effect and correlation with other biomarkers, *Environ Pollut. 156*, 1340-7.

[57] Shinogi, M., Nishinaga, K. and Yokohama, I. (1996) Measurement of metallothionein in rat tissues by a chromatographic method using the fluorimerty of thiol groups, *Biol Pharm Bull. 19*, 911-4.

[58] Imai, K. and Toyo'oka, T. (1987) Fluorometric assay of thiols with fluorobenzoxadiazoles, *Methods Enzymol. 143*, 67-75.

[59] Ip, W. and Fellows, M. E. (1990) Fluorescent measurement of desmin intermediate filament assembly, *Anal Biochem. 185*, 10-6.

[60] Kosower, N. S. and Kosower, E. M. (1987) Thiol labeling with bromobimanes, *Methods Enzymol. 143*, 76-84.

[61] Miyairi, S., Shibata, S. and Naganuma, A. (1998) Determination of metallothionein by high-performance liquid chromatography with fluorescence detection using an isocratic solvent system, *Anal Biochem. 258*, 168-75.

[62] O'Keefe, D. O. (1994) Quantitative electrophoretic analysis of proteins labeled with monobromobimane, *Anal Biochem. 222*, 86-94.

[63] Meloni, G., Knipp, M. and Vasak, M. (2005) Detection of neuronal growth inhibitory factor (metallothionein-3) in polyacrylamide gels and by Western blot analysis, *J Biochem Biophys Methods. 64*, 76-81.

[64] Chinn, P. C., Pigiet, V. and Fahey, R. C. (1986) Determination of thiol proteins using monobromobimane labeling and high-performance liquid chromatographic analysis: application to *Escherichia coli* thioredoxin, *Anal Biochem. 159*, 143-9.

[65] Hamza-Chaffai, A., Amiard, J. C., Pellerin, J., Joux, L. and Berthet, B. (2000) The potential use of metallothionein in the clam *Ruditapes decussatus* as a biomarker of in situ metal exposure, *Comp Biochem Physiol C. 127*, 185-97.

[66] Geret, F., Rainglet, F. and Cosson, R. P. (1998) Comparison between isolation protocols commonly used for the purification of mollusc metallothioneins, *Mar Environ Res. 46*, 545-50.

[67] Jebali, J., Banni, M., Gerbej, H., Boussetta, H., Lopez-Barea, J. and Alhama, J. (2008) Metallothionein induction by Cu, Cd and Hg in *Dicentrarchus labrax* liver: assessment by RP-HPLC with fluorescence detection and spectrophotometry, *Mar Environ Res. 65*, 358-63.

[68] Rodriguez-Ortega, M. J., Alhama, J., Funes, V., Romero-Ruiz, A., Rodriguez-Ariza, A. and Lopez-Barea, J. (2002) Biochemical biomarkers of pollution in the clam *Chamaelea gallina* from south-Spanish littoral, *Environ Toxicol Chem. 21*, 542-9.

[69] Jebali, J., Banni, M., Gerbej, H., Almeida, E. A., Bannaoui, A. and Boussetta, H. (2006) Effects of malathion and cadmium on acetylcholinesterase activity and metallothionein levels in the fish *Seriola dumerilli*, *Fish Physiol Biochem. 32*, 93-8.

[70] Viarengo, A., Ponzano, E., Dondero, F. and Fabbri, R. (1997) A simple spectrophotometric method for metallothionein evaluation in marine organisms: an application to Mediterranean and Antarctic molluscs, *Mar Environ Res. 44*, 69-84.

[71] Bourdineaud, J. P., Baudrimont, M., Gonzalez, P. and Moreau, J. L. (2006) Challenging the model for induction of metallothionein gene expression, *Biochimie. 88*, 1787-92.

[72] Hidalgo, J. and Flos, R. (1986) Dogfish metallothionein-II. Electrophoretic studies and comparison with rat metallothionein, *Comp Biochem Physiol C. 83*, 105-9.

[73] Hidalgo, J., Bernues, J., Thomas, D. G. and Garvey, J. S. (1988) Effect of 2-mercaptoethanol on the electrophoretic behavior of rat and dogfish metallothionein and chromatographic evidence of a naturally occurring metallothionein polymerization, *Comp Biochem Physiol C. 89*, 191-6.

[74] Aoki, Y., Tohyama, C. and Suzuki, K. T. (1991) A western blotting procedure for detection of metallothionein, *J Biochem Biophys Methods. 23*, 207-16.

[75] Haase, H. and Maret, W. (2008) Partial oxidation and oxidative polymerization of metallothionein, *Electrophoresis. 29*, 4169-76.

[76] Kimura, M., Koizumi, S. and Otsuka, F. (1991) Detection of carboxy-methylmetallothionein by sodium dodecyl sulfate-polyacrylamide gel electrophoresis, *Methods Enzymol. 205*, 114-9.

[77] Edwards, R. A. and Maloy, S. R. (2001) Protein aggregation mediated by cysteine oxidation during the stacking phase of discontinuous buffer SDS-PAGE, *BioTechiques. 30*, 311-6.

[78] Hylland, K., Haux, C. and Hogstrand, C. (1995) Immunological characterization of metallothionein in marine and freshwater fish, *Mar Environ Res. 39*, 111-5.

[79] Romero-Isart, N. and Vasak, M. (2002) Advances in the structure and chemistry of metallothioneins, *J Inorg Biochem. 88*, 388-96.

[80] Moreau, J. L., Baudrimont, M., Carrier, P., Peltier, G. and Bourdineaud, J. P. (2008) Metal binding and antioxidant properties of chimeric tri- and tetra-domained metallothioneins, *Biochimie. 90*, 705-16.

[81] Krezel, A. and Maret, W. (2007) Different redox states of metallothionein/thionein in biological tissue, *Biochem J. 402*, 551-8.

[82] Krezel, A., Hao, Q. and Maret, W. (2007) The zinc/thiolate redox biochemistry of metallothionein and the control of zinc ion fluctuations in cell signaling, *Arch Biochem Biophys. 463*, 188-200.

[83] Geret, F. and Cosson, R. P. (2002) Induction of specific isoforms of metallothionein in mussel tissues after exposure to cadmium or mercury, *Arch Environ Contam Toxicol. 42*, 36-42.

[84] Schagger, H. and von Jagow, G. (1987) Tricine-sodium dodecyl sulfate-polyacrylamide gel electrophoresis for the separation of proteins in the range from 1 to 100 kDa, *Anal Biochem. 166*, 368-79.

[85] Urwin, V. E. and Jackson, P. (1993) Two-dimensional polyacrylamide gel electrophoresis of proteins labeled with the fluorophore monobromobimane prior to first-dimensional isoelectric focusing: imaging of the fluorescent protein spot patterns using a cooled charge-coupled device, *Anal Biochem. 209*, 57-62.

[86] Zhang, H., Lian, C. and Shen, Z. (2009) Proteomic identification of small, copper-responsive proteins in germinating embryos of *Oryza sativa*, *Ann Bot (Lond). 103*, 923-30.

[87] Wang, R., Sens, D. A., Garrett, S., Somji, S., Sens, M. A. and Lu, X. (2007) The resistance of metallothionein to proteolytic digestion: an LC-MS/MS analysis, *Electrophoresis. 28*, 2942-52.

[88] Alkhalfioui, F., Renard, M., Vensel, W. H., Wong, J., Tanaka, C. K., Hurkman, W. J., Buchanan, B. B. and Montrichard, F. (2007) Thioredoxin-linked proteins are reduced during germination of *Medicago truncatula* seeds, *Plant Physiol. 144*, 1559-79.

[89] Wong, J. H., Cai, N., Balmer, Y., Tanaka, C. K., Vensel, W. H., Hurkman, W. J. and Buchanan, B. B. (2004) Thioredoxin targets of developing wheat seeds identified by complementary proteomic approaches, *Phytochemistry. 65*, 1629-40.

In: New Developments in Biodiversity … ISBN: 978-1-61324-374-9
Editor: Thomas W. Pace

Chapter 6

LINKING THE DYNAMICS OF RUMINANT FEEDING BEHAVIOR AND DOMINANT SHRUB RESPONSES ON RANGELAND: FORAGE RESOURCES RENEWAL AND BIODIVERSITY CONSERVATION

Cyril Agreil[1*], Danièle Magda[2], Michel Meuret[1], Laurent Hazard[2] and Pierre-Louis Osty[2]

[1]INRA UR767. Sad Ecodéveloppement. Agroparc. F-84 914 Avignon Cedex 9, France

[2]INRA UMR1248. AGIR Sad Orphée. BP 52626. F-31 326 Castanet-Tolosan, France

ABSTRACT

After several decades of marginalization within farming systems, European rangelands are now being challenged to contribute to the conservation of ecological habitats and biodiversity. One of the main challenges, supported by the European Union incentives, relates to the reconciliation of livestock farmers' grazing practices and to the control of dominant plant dynamics, especially those of shrub species, which

[*] Corresponding author: Cyril Agreil, INRA UR767. Sad Ecodéveloppement. Agroparc. F 84 914 Avignon Cedex 9. FRANCE. Email: agreil@avignon.inra.fr

includes maintaining them at density levels appropriate for both habitat conservation and forage resources production. In this chapter, we aim to identify reasons for the difficulty in designing relevant management practices, with focus on the interlinkage of knowledge produced by animal sciences and plant population ecology. From the point of view of these two disciplines, we stress the importance of taking into account the reciprocal interactions between ruminants' foraging strategy and shrubs' demographic behavior. A series of results is given for our experiments on rangelands encroached by Scotch Broom shrubs (Cytisus scoparius L.Linck) and grazed by ewes. Considering the dynamics of ewes' behavioral patterns, we argue for a description of heterogeneous vegetation that recognizes feed items and their functionality for ruminants in maintaining their intake level in a fenced paddock. And considering the dynamic response of Scotch Broom shrubs to browsing, we also argue for demographic models that would recognize the importance of both the selective offtake of plant organs and the new browsing-induced demographic behavior of shrubs. These results enabled to identify the plant community as a mandatory intermediate object, and the plant organs as a key organization level at which these two processes interact. We propose an original conceptual framework that interlinks the two processes and recognizes the specific organizational levels and time frames. This framework should facilitate the identification of prospective research issues such as the differential impact of browsing on shrub demography according to plant organs and life stages consumed, or the effect of different vegetation states of a plant community on selective browsing among shrub organs. For rangeland management, the framework brings out the importance of greater precision in identifying the targets, and in particular the target plant organs and target life stages in shrub demography control. Considering this objective the choice of the season for grazing a given fenced pasture should also be made bearing in mind the global feeding offer within the plant community.

Keywords: browsing, small ruminants, shrub population, demography, dominant species, foraging

INTRODUCTION

In Europe, "rangelands" are defined as rural areas bearing semi-natural vegetation that have been devoted for centuries to pastoral husbandry, mainly sheepherding. Rangelands may be located in the high mountain summer, or in the plains and the foothills where they constitute an intermediate space between croplands and forests. In contrast to cultivated grasslands, livestock farmers manage their herds or flocks on rangelands in such a way as to ensure that the animals are adequately fed while simultaneously seeing to it that the land's forage resources will be renewed and available for future seasons or years. They achieve that through herding and/or rotational grazing in rather small fenced pastures, i.e. a few acres. Both techniques are used to apply high-grazing pressure locally and for a short period of time, once or twice a year. In some situations, the farmer may supplement the effect of grazing with controlled burning operations or the use of machinery to cut and slash excessively dense brush.

European rangelands were faced with two radical changes (Hubert *et al.*, *in press*). Until the second half of the 19th century, they were an integral part of mixed farming production systems, used mainly as a fertility reserve for croplands for a fast-growing rural population. Large shrubs, such as broom species, were used for farm and shed roofing or for fuel in bakers' ovens. Throughout the 20th century, as urban areas were becoming predominant, rangelands were marginalized by the widespread adoption of "rational" agricultural techniques. They served no purpose for livestock farmers that concentrates their activities in the plains and valley bottoms, where they fed their new breeds of demanding animals from cultivated grasslands and concentrated feed. Most rangelands were abandoned or planted with trees and the brush proliferation and afforestation process was dynamic (see Picture 1): brush and tree cover expanded by 50 to 60% in southern France over a period of 40 years. In the 1980s, this process was considered as landscape degradation, witness the wildfires problem. Since the 1990s, rangelands have attracted renewed social and political interest because of their environmental goods. Many of these lands have acquired a priority status for their contribution to the conservation of outstanding ecological habitats and threatened wildlife species, pursuant to the European Union's "Habitats Directive", issued in 1992, the year of the Rio Earth Summit (Pinton *et al.*, 2006).

The environmental public policy is expecting much of the remaining farmers as owners or tenants who still use rangelands for grazing. The EU is offering substantial incentives through 5-year grazing contracts for specific parcels on farmlands. The aim is to design a "yearly grazing schedule" for each parcel or group of parcels, in order to more effectively control certain fast-growing and dominant plant dynamics through grazing (Léger *et al.*, 1999).

Picture 1. Proliferation of Cytisus scoparius L.Linck on a foothill in southern France.

Despite the financial rewards, the majority of livestock farmers are unenthusiastic. After decades of "rational" feeding techniques, they are now supposed to turn their animals to graze "unknown territory", since virtually none of the standard reference works on livestock feeding deals with rangelands, and the memory of the way such lands were formerly used has been lost. What sort of edible forage plants grow there? What is their feeding value? Is grazing a reliable technique for slowing down the dynamics of dominant shrubs, and restoring or maintaining the forage resource quality? Facing such a great lack of knowledge, most of the farmers prefer to eradicate brush by machine, winter burning or chemical product.

The environmental policy do not support this type of shrubs and trees eradication since its aim is to preserve a certain amount of them on each parcel as a functional component of diversified plant communities and wildlife habitats. The management of the dynamics of some dominant shrubs thus is, in itself, a major goal for ecological quality conservation and of forage resources renewal. For livestock farmers to consider edible shrubs as a component of the pasture would be a major breakthrough in Europe and elsewhere, both in the farmers' mindsets and in pasture management techniques. Flock management practices that could drive or at least slow down the dominance dynamics of shrubs would have to be identified or created. The stakes are high. Animal and plant scientists should be involved to provide knowledge to renew grazing practices.

From a scientific standpoint, little is known about the dynamic responses of dominant shrub species to different grazing management regimes. Research on the interactions between animal grazing behavior and plant dynamics (refer to Papachristou *et al.*, 2005; Shipley, 2007 for reviews) is still incomplete. But most important, the produced knowledge rarely examines animal foraging behavior and plant population dynamics as two interdependent, adaptive processes. Many papers have dealt with plant dynamics in response to grazing, but only considered grazing as a static input (Ammer, 1996; Bellingham and Allan, 2003; Hübler *et al.*, 2005). Others have focused on the dynamics of feeding behavior, and characterize plant cover by static vegetation states (Bergman *et al.*, 2005; Bergvall *et al.*, 2006).

Some recent developments have tried to link the two processes but only focused on one aspect of their dynamic properties, without really modeling the loops of their reciprocal dynamic responses. Danell *et al.* (2003), for instance, looked at the impact of selective herbivory on morphological changes and on the density of adult trees of a species, but did not characterize changes in the structure and demography of the plant population nor the subsequent herbivore feeding behavior in response to these changes. Cooper et *al.* (2003) described these reciprocal dynamic responses through changes in shrub architecture, but made no reference to the impact on the shrub population demography.

It is therefore crucial to progress towards a synthetic representation of these two processes as constituent elements of a dynamic system. And for science, the challenge cannot be limited to an innovative interlinking of existing knowledge, emphasis must also be placed on the imperative need to produce knowledge for characterizing the specific dynamics of both foraging behavior and shrub population dynamics. Section 1 will focus on the foraging behavior of ruminants when feeding on heterogeneous vegetation, and Section 2 will focus on modeling the responses of shrub populations demography to herbivory. We will illustrate how our recent experimental results can complement existing knowledge in animal science and plant ecology, and pave the way to a dynamic interlinking of ruminant foraging and dominant shrub dynamics. On the basis of these findings, Section 3 presents an original conceptual framework is presented that accommodates the nature of the interactions between these two processes, viewed as dynamic, complex and interdependent. In the conclusive section we discuss the aforementioned framework as a first step of a cognitive process that makes it possible to identify the research perspectives that need to be investigated, and to revise the design for the management plans.

1. The Feeding Behavior of Ruminants as an Adaptive Process: Redefining the Foraging Strategy

Up to now, animal sciences have tended to produce knowledge and feeding standards designed for homogeneous or cultivated pastures (Wilson *et al.*, 1995). For rangelands, where vegetation is particularly diversified, this often leads to an inaccurate description of plant-animal grazing interactions, either considering the grazing process as a homogeneous offtake of the edible plant material or considering the feeding behavior to be largely determined by intrinsic plant characteristics.

However, when associated to foraging ecology, animal sciences have produced some knowledge that brought out the dynamics of foraging behavior. Domestic ruminants faced with heterogeneous forage resources can now be considered in terms of the behavioral adjustments that constitute feeding strategies for meeting their nutritional requirements (Laca and Demment, 1996). They adjust their qualitative and quantitative uptake in response to the plants' architecture and biochemical composition, which also vary over space and time. A major challenge for foraging ecology is to advance understanding of the roles played by different components of vegetation diversity (described by species, organs, architecture, biochemical composition, etc.) within these feeding strategies (Provenza *et al.*, 2003; Agreil *et al.*, 2006).

On rangelands, ruminants have to choose their feed not only among species, but also among plant organs within the species, e.g. twigs, leaves and fruits (O'Reagain and Grau 1995; Gillingham *et al.*, 1997; Agreil and Meuret, 2004). Their prioritization criteria then result in temporal patterns at different temporal scales, including the situations in which the feed offer remains unchanged in time. Ruminants' choice criteria vary, for instance, within a meal (Rolls 1986, Gillingham and Bunnel 1989, Newman *et al.*, 1992), within a day or between days, depending on their basal diet (Parsons *et al.*, 1994). When the feed offer varies over time, ruminants are not only subordinated by the temporal variation of vegetation, they also actively adjust their feeding behavior and choice criteria. When grazed in successive fenced pastures for instance, ruminants select species sequentially in response to the major changes provoked by offtake (O'Reagain and Grau 1995).

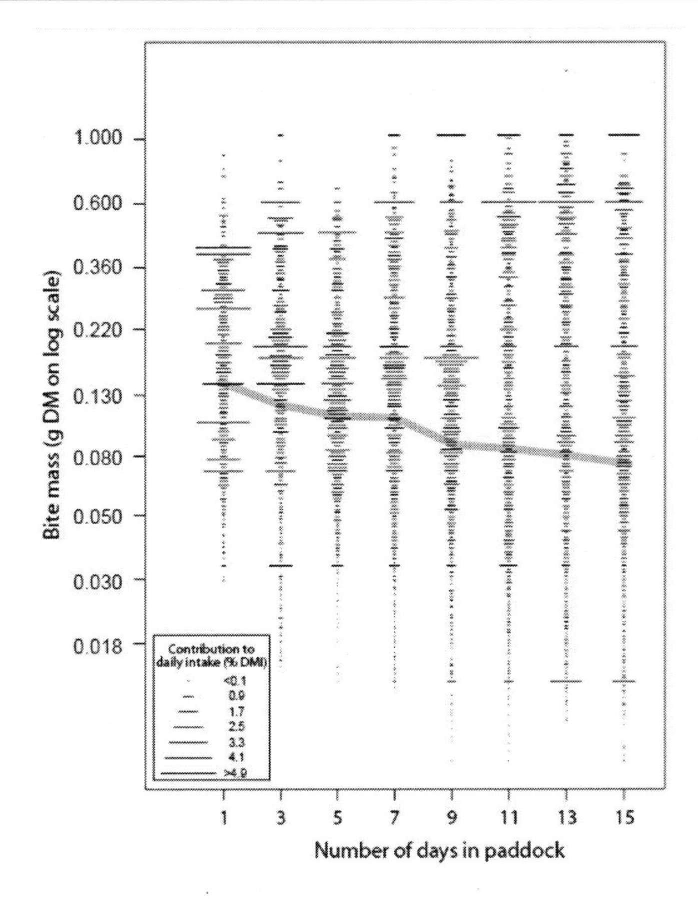

Figure 1. Inter-day variation of the distribution of bite masses during the fenced pasture utilization sequence. For each category of bite mass, the length of the black dashes is proportionate to the average contribution of the bite mass category to the total daily intake of dry matter (%).

Since not enough effort has been made yet to describe the variation of feeding behavior over time, results between different experimental scales and contexts may seem to be contradictory. The many debates concerning the currencies and functions to be used in foraging models are very insightful (Owen-Smith N. 1993, Belovsky and Schmitz 1993, Sih and Christensen 2001). The ability to record and analyze feeding behavior, taking into accounts its heterogeneity and temporal variability is central to any study on feeding choices of ruminants' faced with a highly diversified feeding offer. Techniques for direct observation and continuous recording of bites (Stobbs, 1975, Meuret *et al.*, 1985) have been recently improved (Parker *et al.*, 1993; Agreil and

Meuret, 2004), and now allow for simultaneous estimates of mid- and long-term intakes, and the compilation of feeding choice sequences that underlie these intakes (Gilligham *et al.*, 1997; Mofareh *et al.*, 1997; Kohlmann *et al.*, 1999). In the next part of this section, we present detailed results from comprehensive full-day records of sheep bites records, because these data make it possible to carry out multi-scale analyses of behavioral dynamics (Agreil and Meuret, 2004).

Experiments were conducted with flocks of dry ewes, grazed for 10 to 16 days in small pastures (1-4.5 ha, 2.5-11.1 acres) in southern France. In order to avoid any artifact due to the feeding habits of the flocks, the ewes were observed on farm, within fenced pastures they were used to grazing during the season of study (Agreil *et al.*, 2006). In order to identify generic choice rules among plant organs, we analyzed the behavioral data at several temporal scales. This led to the identification of two major behavioral trends that strongly shape behavioral adjustments for coping with heterogeneous vegetation.

At the scale of a sequence of days spent within a fenced pasture, as resources grew scarce and were finally depleted over a period of days, ewes not only shifted to taking smaller bites on previously grazed plants but also gradually shifted to taking larger bites from plant organs not previously selected (Figure 1), thus stabilizing their average intake rate per meal (Agreil *et al.*, 2005). Late selection of large bite masses (>0.15 g DM/per bite or >50 mg DM/kg $LW^{0.75}$) contributed to maintaining the stability of daily intake until nearly the end of the animals' period of stay within the fenced pasture (see the thick gray line on Figure 1). This behavioral trend led to a variable use of plant organs within species. For a shrubby species like broom, the temporal variation of choice criteria meant prior consumption of highly palatable bites mixing pods and twigs (Figure 2). As the days went by, the ewes gradually shifted to large bites and hence increased their consumption of large bites on broom long twigs (Figure 2).

A second important temporal scale is the meal. At this scale, the ewes under observation consumed a broad range of edible plant organs, including those of shrub species (Figure 3). The temporal succession of bites is amazingly diversified in terms of mass, species and biochemical composition. By modeling temporal dependency within sequences of intake rates during meals, we detected an oscillating dependency on the past, which indicates that intake rate oscillated over and under a long-term trend line (Agreil *et al.*, *submitted*). At the meal scale, bite mass also appears to be a relevant decisional cue indicating how small ruminants organize their feeding strategy and maintain their intake efficiency at a level that can cover the dietary requirements.

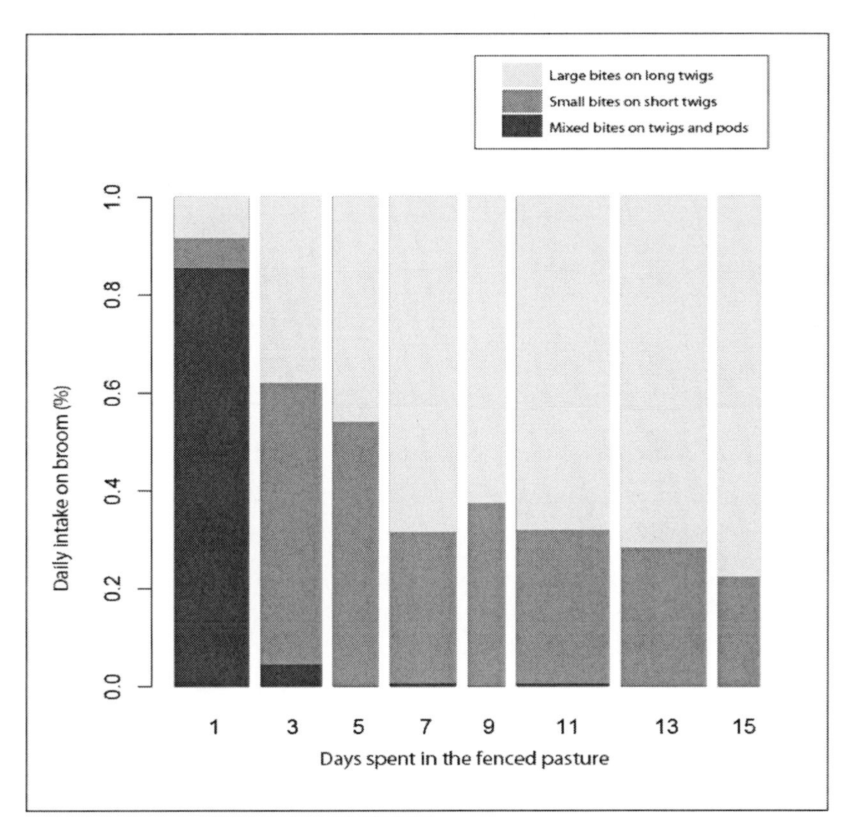

Figure 2. Evolution of the plant organs (large bites on long twigs in light gray, small bites on short twigs in intermediate gray and mixed bites on twigs and pods in dark gray) selected by ewes on Scotch Broom (*Cytisus scoparius* L.Linck) over the days spent in a fenced pasture. Bar width represents the cumulated broom dry matter ingested each day.

These results encourage us to stress the importance of considering the feeding behavior of ruminants as a dynamic, adaptive process, leading to different but ultimately predictable diets as a function of the given temporal scales and the feeding offer composition. Since modeling dominant shrub species is the topic of this chapter, we also want to emphasize models that recognize foraging behavior in relation to the global composition of the plant community. But in order to understand or predict foraging choices, the description of the plant community must also include the functionality of the various edible plant organs, according to their structure, that determines whether the ruminants will take either large or small bite masses. This structural diversity is found not only between species but also within species, e.g. between the different organs of a shrubby plant. The early results presented here introduce a way to update the characterization of vegetation through a functional description, which will be of great help in interlinking foraging behavior and a dominant shrub species.

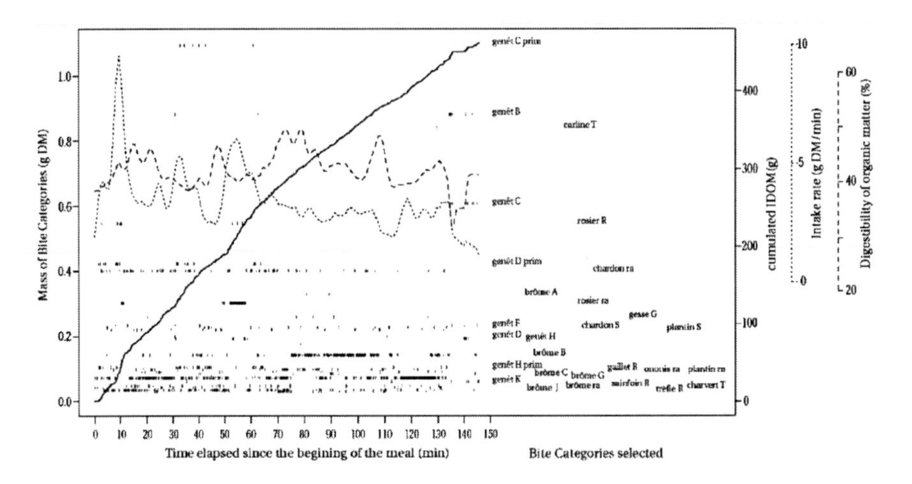

Figure 3. Profile of a meal observed on the first day of utilization of a fenced pasture. The figure gives a chronicle of bite category selection: each little vertical bar represents one bite. Its position in relation to the vertical axis on the left gives information on its mass (DM, in g). For the 28 bite categories most recorded, the name of the plant followed by the code for the plant organ selected are given on the right (see Agreil and Meuret, 2004 for details). The effects of these choices on cumulated ingested digestible organic matter (IDOM, in g: vertical axis on the right) are shown on the solid line. The effects of these choices on changes in the DM intake rate (in g $DM.min^{-1}$, fine dotted line) and the organic matter digestibility (OMd in %, large dotted line) are also given.

2. Shrub Population Dynamics in Response to Animal Feeding Behavior: New Developments for Demography Modeling

Plant population ecology has developed complex demographic models for several species (Florian *et al.*, 2007). Most of them are poorly predictive when the population is subject to herbivory since they ignore the feeding behavior of the herbivores (from insects to mammals). These models generally estimate the impact of herbivores through an indirect estimation of biomass removal on plant organs at a given time, and consider this impact as a constant variation in one or more demographic parameters over the years (Leimu *et al.*, 2006; Kelly *et al.*, 2002; Doak, 1992). For the case of shrub browsing in rangelands, these models ignore both the intensity variations and the temporal occurrence of the plant organ removal. On rangelands, ruminants can select on shrub species, a great diversity of organs (flowers, fruits, twigs) of various life stages (seedlings, juveniles and adults) (Hansson & Fogelfors, 2000; Valderràbano & Torrano, 2000; Rousset & Lepart, 2002; Frost & Launchbaugh, 2003), thereby suggesting that browsing can affect different demographic processes. The impact of browsing on shrubs has been treated by focusing on a life stage that is considered *a priori* as the main driver for demography regulation but without testing it at the population level (Belligham & Coomes, 2003; Bartolomé *et al.*, 2005; Seifan & Kadmon, 2006). These species are known to develop adaptive morphological or phenological responses when faced with repeated herbivory. The responses developed by these long living shrub species are generally ignored in models, despite their consequences on the plant's demographic behavior.

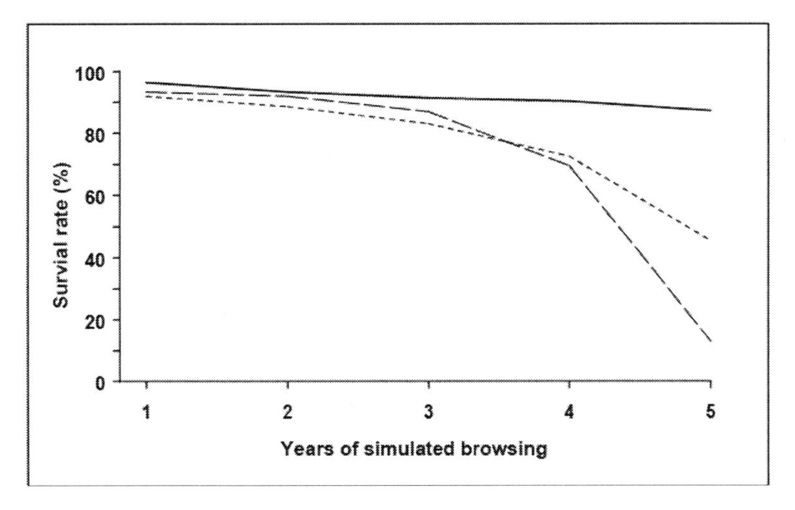

Figure 4. Evolution of survival rate (%) of Scotch Broom (*Cytisus scoparius* L.Linck) juveniles over a period of 5 years, as a function of simulated browsing intensity: control (solid line), low intensity (fine dotted line), and high intensity (large dotted line). Effect of simulated browsing (low and high intensity) is significant as of the fifth year ($\chi^2 = 2788.97$; $P \leq 0.001$).

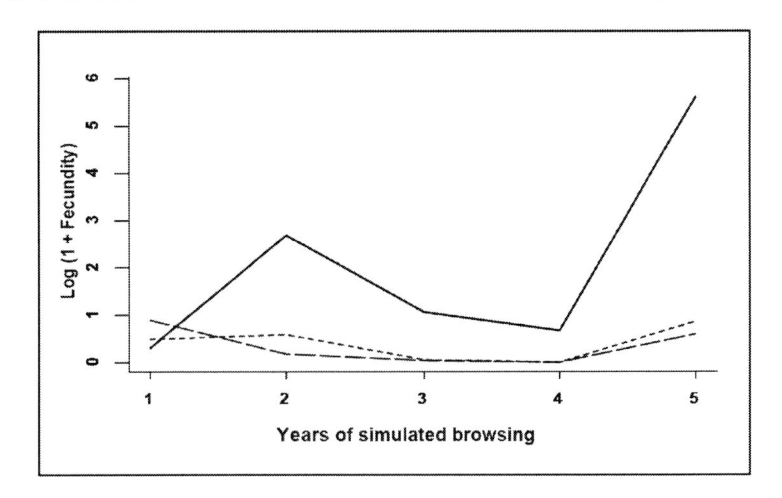

Figure 5. Evolution of mean fecundity (estimated by the mean number of pods per individual) for Scotch Broom (*Cytisus scoparius* L.Linck) juveniles, over a period of 5 years of as a function of the simulated browsing intensity: control, low intensity (fine dotted line), and high intensity (large dotted line). Effect of grazing, compared to control, is significantly different as of the second year (test F, $P \leq 0.001$).

Shrubs may adopt different strategies in response to browsing (Briske, 1999; Shipley, 2007). They may, for example, increase their chemical (e.g. lignin) and physical (e.g. spines and thorns) properties that serve as grazing deterrents (Laca *et al.*, 2001; Papachristou *et al.*, 2005). Repeated consumption of mature twigs can cause lasting or even irreversible effects on the adult phenotypes (Cooper *et al.*, 2003). Shrubs may also favor the increased shoot regrowth ability (Westoby, 1999; Del-Val & Crawley, 2005), like a tolerance strategy to this disturbance, but at the expense of their reproductive organs, which may not develop at all.

To improve predictions on the potential impact of grazing on shrub population will require upgraded demography modeling based on a better understanding of ruminant feeding behavior. First, the model should make it possible to explore the impact of browsing of the different organs and life stages, identified as food items for ruminants, on the population growth rate, in order to suggest which target organs and life stages should be managed. Second, the model should allow for the integration of plant adaptive responses to browsing, which could deeply alter the population's demographic behavior. Using Scotch broom shrub (*Cytisus scoparius* L.Linck) as a model species, we developed an approach based on a demographic shrub analysis designed to explore the diversity of the shrub population responses to grazing and to propose relevant management practices. We started by defining what is called the demographic shrub strategy by analyzing the specific structure of the population in different life stages (seed, seedling juvenile, adult) and quantified all the values of the demographic parameters (seed survival rate, fecundity, adult survival rate) by determining the transition rate between the life stages. With adequate modeling tools, this makes it possible to determine the relative weight of each demographic parameter and each life stage on the population growth process. The most important parameters are identified in the species demography strategy analysis. This allows for the identification of the targeted life stages and targeted plant organs, whose consumption through browsing is expected to have major consequences on population growth process. For Scotch Broom, the analysis showed that the survival rates of seeds, seedlings and juveniles and the young adult fecundity are the parameters that have the most impact on the population growth rate (Magda *et al., submitted*). As seeds and seedlings are not directly accessible life stages for ruminants, this analysis shows that the juveniles and young adults are the appropriate potential target life stages for management. For the impact of browsing to be predicted with precision, quantitative relationships must be developed between the intensity of consumption of the different organs of each life stage and the variations in the demographic parameters values. This issue can be quite complex because of the many processes involving the plant and their relation, albeit indirect, to demography. For Scotch Broom, we showed that consumption of reproductive organs, which directly and proportionately affects fecundity, cannot be used as an effective population density control tool. To stabilize the population size would require quasi-total removal of their reproductive organs by ruminants, every year. These results are at variance with the empirical practices that encourage the grazing of flowers or pods since that is the easiest and more effective way to influence shrub dynamics. The impact of twig browsing on plant survival or fecundity is less easy to quantify. To control Scotch Broom, the most strategic life stage to impact is the juvenile stage but no data are available to estimate the correlation between the importance of twig browsing and juvenile survival. To assess this relationship, we simulated the impact of two different intensities of annual browsing of juvenile twigs on juvenile survival and on the transition rate to reproductive adulthood. We showed that the juvenile survival rate decreases significantly after five years regardless of intensity (Figure 4) and that browsing prevents the transition to reproductive adulthood (Figure 5). These results show that browsing can strongly affect the demographic behavior of the juvenile life stage, with regard to both survival and reproduction processes.

Juvenile response to repeated browsing leads to a reallocation of resources to vegetative growth. This juvenilization process creates a new category in the population that we named "juvenilized adults", a browsing-induced category whose demographic behavior is radically different from that of the reproductive young adults. In order to take account of the long-term impact of browsing, we added this new category to the population structure (Figure 6). Shrub population responses to browsing are complex because of the diversity of plant organs that ruminants can consume. The nature and time scale of responses depend upon the plant development process and can vary greatly. This knowledge is essential in modeling a representation of the population which will be relevant in creating the link with the feeding behavior process.

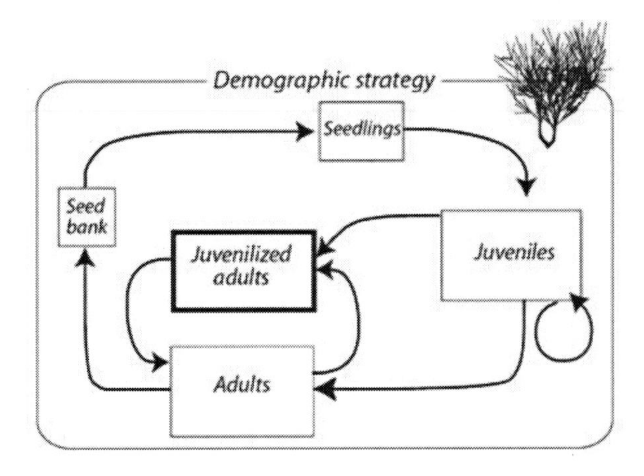

Figure 6. The life cycle of Scotch Broom (*Cytisus scoparius* L.Linck) represented by a life stage structured model with the four main stages: seeds in the seed bank, seedlings, juveniles, and reproductive adults. The juvenilzed category is added as a new category induced by the response of juveniles and adults to repeated browsing.

3. Interlinking the Feeding Behavior of a Small Ruminant and the Population Dynamics of a Dominant Shrub

These two research fields have enabled to identify the main requirements for revisiting the concept of the interlinkage of ruminants' foraging behavior and dominant shrubs dynamics. First and foremost we recognized that the description of the intrinsic dynamics of the two biological processes had to be clear before we could interlink the two processes functionally. In a first but crucial stage, this knowledge enabled us to identify the main scales and the corresponding organizational levels at which the two processes interact.

A schematic representation of our proposal for a conceptual framework of these interactions is given in a schematic representation (Figure 7) showing four temporal scales (vertical columns: year, fenced pasture utilization period, day, and meal) and three biological objects (horizontal rows: animals, plant community and dominant shrub population).

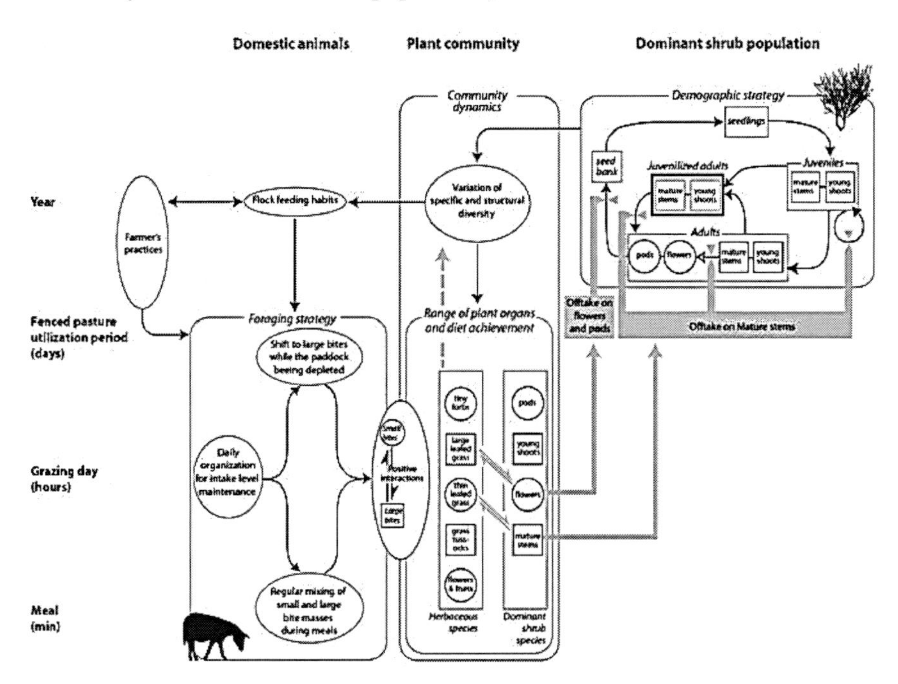

Figure 7. A conceptual framework to link the feeding strategy of a small ruminant and the population dynamics of a dominant shrub species within a plant community. Four temporal scales are distinguished and labeled (left). Three processes related respectively to animals, plant community and dominant shrub population are explicitly described in boxes. The functional links between the foraging strategy of a small ruminant (bottom left) and the population dynamics of a dominant shrub species (top right) can be understood provided that: 1) the range of edible plant organs available in the community is described in terms of functional feeds; 2) the consequences of selective offtake of shrub organs by animals are described in terms of changes in demographic behavior. Grey arrows show, for two examples, the different browsing-induced impacts on shrub demographic behavior according to the nature of plant organs and of life stages browsed.

The most central attribute of the framework is the key status we assigned to the plant organs, as the main organizational level of the interaction between animal foraging and plant population dynamics. This organizational level determines the choice distribution in the feeding offer and the effects of selective browsing on shrub demographic behavior. The second main attribute of the framework is that it considers the interaction between ruminants and a dominant shrub population via an intermediate object: the plant community. This framework, thus, recognizes that ruminants develop their feeding strategies in relation to the feeding offer within the plant community as a whole, and not only in relation to the shrubby species, even if they are dominant. The global description of the plant community should be useful in anticipating the foraging behavior and the distribution of choices, especially within the shrub population, provided we are able to specify the particular roles of the different components (from herbaceous cover and shrub species) within the foraging strategy. In the case of broom shrubland, for instance, the abundance of large leafed grasses within the herbaceous cover, which are harvested through large bites, will probably motivate ruminants to select small plant organs, in particular flowers or pods on the dominant shrub species (see thick gray arrows on Figure 7).

Concerning the dominant shrub species, the distinction of plant organs within the dominant shrub population is of great importance in assessing the consequences of selective foraging. With this in mind, the framework includes the grazing-induced categories in the life cycle plus the availability of the different plant organs provided in each demographic category, thus making it possible to account for the different impacts on demographic parameters of the browsing of different plant organs. The browsing of pods or flowers, for instance, directly impacts fecundity, whereas the browsing of stems impacts the survival rate of the individual (see thick gray arrows on Figure 7). This also accounts for the fact that the removal of a given plant organ will have consequences on different demographic parameters depending on the development stage. The removal of mature twigs, for instance, (see thick gray arrows on Figure 7) could decrease the juvenile survival rate (if removed from juveniles), decrease the transition towards reproductive adults (if removed from juvenilized adults), or decrease flowers and pods production (if removed from adults).

Finally, another important attribute of the framework is its recognition that the farmers' management practices focus on the flocks and not on the shrub's demographic dynamics. His pasturing rules gradually shape the flocks' feeding habits, which in turn largely determine individual foraging decisions (see Management perspectives section below).

CONCLUSION

Thanks to this conceptual framework, we can now propose a series of perspectives and requirements for the study, modeling and management of vegetation dynamics on grazed rangelands.

First, the framework stresses the importance of recognizing the intensity and distribution of browsing impacts not only within the demographic plant categories, but also among their plant organs. There is an obvious need for research on the precise consequences of such differentiated browsing impact, in particular by identifying the time frames for shrubs demographic behavioral changes. This research should also serve to identify the target demographic life stages, and the target plant organs as well as processes that have to be impacted in order to efficiently orient the demographic behavior of the population.

Second, the framework stresses the importance of recognizing that ruminants' browsing impact on shrubs largely depends on the range and abundance of plant organs within the plant community. The maturation stage and height of the herbaceous cover, for instance, is likely to influence the distribution and intensity of browsing among the available plant organs on shrubs. In the long term, how the grazing management regimes influence the resulting assemblage of plant organs within the community is difficult to predict. This assemblage depends on several processes that occur at different levels in the community organization: plant development, plant phenology, population dynamics, interspecies competition, etc. Data are not yet available on the relationships between this assemblage and the species diversity composition comprising the dominant shrub species. Developing the very few existing scientific studies on this topic would contribute to identifying which herbaceous states encourage ruminants to browse some plant organs rather than others. As an output (see next below), management plans could be designed to support a particular target season.

Third, the framework has been based on the recognition of respective process dynamics, which meant building interlinkages with due attention to their reciprocal dynamic interactions. Thanks to its design, the framework is useful in examining how the foraging behavior impacts the dominant shrub demography, which in turn provokes changes in the feed offered by the plant community, which in turn leads to further adjustments in foraging behavior. But yet, the framework has not been built to study the mechanisms that condition changes within the plant community. Were this attribute to be added, the feedback loop could be more precisely specified, and the subsequent behavior predicted.

In terms of management, the first and main implication is probably to recognize the complexity of the functional links between management practices and the resulting control of a dominant shrub species. When management is principally based on flock handling and tends to exclude mechanical clearance or burning, many biological processes interact, *in fine*, to orient vegetation dynamics. Our framework should facilitate the interpretation of different vegetation states obtained through seemingly similar management practices, because it precludes the overly simplified correlation between management input and vegetation state output. This should, in turn, be useful in defining the management indicators needed to make a diagnostic of the feeding offer and of the impact on shrubs.

A second management implication is the importance to relate management practices to observations of the life stages of pasture vegetation and the edible plant organs. These relevant life stages and plant organs are, of course, functional feeds for the flock, but may also be considered as grazing targets if identified as decisive in controlling the shrub demography strategy. Should this targeted grazing be successful, density levels that are compatible with forage resources and ecological quality maintenance could be achieved without requiring any major interventions (see Picture 2).

And finally, a third important management implication is the choice of season, which should be aligned to the animals' attraction to the targeted plant organs. From this vantage point, the fenced pasture utilization period (season, duration) is essential. Through his choice of season for pasture grazing, the livestock farmer decides on the vegetation state at the time of entrance, and hence decides to put the flock in a plant community with a given structural and phenotypic state, characterized by certain assemblages of plant organs. Knowing the time of year the various target plant organs are produced and palatable would help the livestock farmer decide on the dates for the grazing season. But it is also important to consider the phenology of the other species that could encourage or dissuade animals to impact the target plant organs. The diagnostic of functional feeds available at the beginning of and during the utilization period of a pasture would thus enable the livestock farmer to schedule and adjust the length of stay in the pasture so as to ensure continuous satisfactory intake levels. Through his choice of an end date for a pasturing period, he decides the exit vegetation state, which is the key criterion in anticipating the subsequent shrub demographic responses. At the end of each utilization period of a fenced pasture, the diagnostic should be focused on the consumption rate of targeted life stages and targeted plant organs, as a prediction of the impact of grazing on the dynamics of the dominant species.

REFERENCES

[1] Ammer C. (1996) Impact of ungulates on structure and dynamics of natural regeneration of mixed mountain forests in the Bavarian Alps. *Forest Ecology and Management*, 88, 43-53.

[2] Bartolomé, J., Lopez, Z. G., Broncano, M. J. and Plaixats, J. (2005) Grassland colonization by E*rica scoparia* (L.) in the Montseny Biosphere Reserve (Spain) after land-use changes. *Agriculture, Ecosystems & Envrionment, 111*, 253-260.

[3] Bellingham P. J. and Allan C. N. (2003) forest regeneration and the influences of white tailed dear (*Odocoileus virginianus*) in cool temprate New Zealanf rain forests. *Forest Ecology and Management*, 175, 71-86.

[4] Bellingham, P. J. and Coomes, D. A. (2003). Grazing and community structure as determinants of invasion success by Scotch broom in a New Zealand montane shrubland. *Diversity and distributions, 9*, 19-28.

[5] Bergman M., Iason G. R., Hester A. J. (2005). Feeding patterns by roe deer and rabbits on pine, willow and birch in relation to spatial arrangement. *Oikos*, 109, 513-520.

[6] Bergvall, U. A., Rautio, P., Kesti, K., Tuomi, J. and Leimar, O. (2006). Associational effects of plant defences in relation to withinand between-patch food choice by a mammalian herbivore: neighbour contrast susceptibility and defence. *Oecologia*, *147*, 253-260.

[7] Briske, D. D. (1996). Strategies of plant survival in grazed systems: a functional interpretation. In: Editors Hodgson J. and Illius A. W. *Ecology and Management of grazing systems*. NewYork: Cab International N. Y. p. 37-68.

[8] Cooper, S. M., Owens, M. K., Spalinger, D. E. and Ginnettn, T. F. (2003). The architecture of shrubs after defoliation and the subsequent feeding behavior of browsers. *Oikos, 100*, 387-393.

[9] Doak, D. F. (1992). Lifetime impacts of herbivory for a perennial plant. *Ecology, 73*, 2086-2099.

[10] Danell, K., Bergström, R., Edenius, L. and Ericsson, G. (2003). Ungulates as drivers of tree population dynamics at module and genet levels. *Forest Ecology and Management, 181*, 67-76.

[11] Florian, J., Moloney, K. A., Schurr, F. M., Köchy, M. and Schwager, M. (2008). The state of plant population modelling in light of environmental change. *Perspectives in Plant Ecology, Evolution and Systematics, 9*, 171-189.

[12] Frost, R. A. and Launchbaugh, K. L. (2003). Prescription Grazing for Wildland Weed Management. A new look at an old tool to control weeds on rangelands. *Rangeland, 25*, 43-47.

[13] Hansson, M. and Fogelfors, H. (2000). Management of a semi-natural grassland; results from a 15-year-old experiment in southern Sweden. *Journal of Vegetation Science, 11*, 31-38.

[14] Hubert, B., Deverre, C. and Meuret, M. (2009). The rangelands of southern France: two centuries of radical change. In: Editors Meuret M., Provenza F. *Learning from Sheepherders*. Versailles, France : Quae;.

[15] Hülber K., Ertl S., Gottfried M., ReiterK., Grabherr G. (2005). Gourmets or gourmands? Diet selection by large ungulates in high-alpine plant communities and possible impacts on plant propagation. *Basic and Applied Ecology*, 6, 1-10.

[16] Kelly, C. A. and Dyer, R. (2002). Demographic consequences of inflorescence-feeding insects for Liatris cylindraceae, an iteroparous perennial. *Oecologia,132*, 350-360.

[17] Laca, E. A., Shipley, L. A., Reid, E. D. (2001). Structural anti-quality characteristics of range and pasture plants. *Journal of Range Management 54*, 413-419.

[18] Léger F. Bellon S. Meuret M. Chabert J. P. Guerin G. (1999) Technical approach to local agro-environmental operations: results and means. In Editors : Rubino R. Morand-Fehr P. *Options Mediterranéennes*, Serie A, 163-167.

[19] Leimu, R. and Lehtilä, K. (2006). Effects of two types of herbivores on the population dynamics of a perennial herb. *Basic and Applied Ecology, 7*, 224-235.

[20] Magda, D., Chambon-Dubreuil, E., Gleizes B. and Jarry M. *Demographic analysis of a dominant shrub* (*Cytisus scoparius*): *prospects for encroachment control by grazing*. Submitted to Basic and Applied Ecology.

[21] Papachristou, T. G., Dziba, L. E. and Provenza, F. D. (2005). Foraging ecology of goats and sheep on wooded rangelands. *Small Ruminant Research, 59*, 141-156.

[22] Pinton, F., Alphandéry, P, Billaud, J. P, Deverre, C., Fortier, A. and Geniaux G., (2006). *La construction du réseau Natura 2000 en France*. Paris : La Documentation Française.

[23] Rousset, O. and Lepart, J. (2002). Neighbourhood effects on the risk of an unpalatable plant being grazed. *Plant Ecology, 165*, 197-206.

[24] Seifan, M. and Kadmon, R. (2006). Indirect effects of cattle grazing on shrub spatial pattern in a Mediterranean scrub community. *Basic and Applied Ecology, 7*, 496-506.

[25] Shipley, L. A. (2007). The influence of bite size on foraging at larger spatial and temporal scales by mammalian herbivores. *Oikos,116*, 1964-1974.

[26] Val, E. del and Crawley, M. J. (2005). Are grazing increaser species better tolerators than decreasers? An experimental assessment of defoliation tolerance in eight British grassland species. *Journal of Ecology, 93,* 1005-1016.

[27] Valderràbano, J. and Torrano, L. (2000). The potential for using goats to control *Genista scorpius* shrubs in European black pine stands. *Forest Ecology and Management, 126*, 377-383.

[28] Westoby, M. A (1999) LHS strategy scheme in relation to grazing and fire. In : Editor D. Eldridge & D. Freudenberger . *Proceedings of the VIth International Rangeland Congress*. Queensland, Australia: Australian Rangeland Society; 893-896.

In: New Developments in Biodiversity ... ISBN: 978-1-61324-374-9
Editor: Thomas W. Pace

Chapter 7

OUTCOMES OF INVASIVE PLANT-NATIVE PLANT INTERACTIONS IN NORTH AMERICAN FRESHWATER WETLANDS: A FOREGONE CONCLUSION?

Catherine A. McGlynn*
Hudsonia Ltd, Red Hook, NY 12571, U.S.A.

ABSTRACT

Freshwater tidal wetlands are productive and often support high biodiversity. While there have been some quantitative studies of the effects of many invasive plants in North American freshwater wetlands, much is still assumed for a number of invasive species. The outcomes of invasive-native plant interactions, and the factors involved, are more subtle than assumed. I reviewed the statistical and anecdotal results of published studies on the impacts of invasive plants with large potential ranges in freshwater tidal wetlands of North America. A number of studies reported little or no change in native species richness and diversity, and found outcomes differed depending upon several specific factors. I recommend that more empirical research be conducted on the interactions of these species (both in field and greenhouse) with native

* Corresponding author: Hudsonia Ltd., P.O. Box 66, Red Hook, NY 12571 Catherine McGlynn
Email: catherine_mcglynn@yahoo.com

plants in freshwater tidal wetlands because very little information is available and yet many management decisions have already been made.

INTRODUCTION

Biological invasions are responsible for billions of dollars in economic losses and an untold number of ecological impacts throughout the world (Mack et al. 2000). In the United States alien species are the second highest threat, after habitat loss and deterioration, to endangered or threatened native species (Wilcove et al. 1998). Along with changes in land-use and climate, biological invasions are drivers for environmental change on a global scale. In aquatic systems, however, the effects of biological invasions often remain unquantified.

Wetland plants (floating, submerged, and emergent) are a vital part of the aquatic ecosystem. They provide habitat and food for many taxa, both vertebrate and invertebrate; and contribute to nutrient cycling and many other ecological functions (Carpenter and Lodge 1986). Living in the diverse and dynamic habitat of freshwater tidal wetlands, wetland plants are subjected to considerable natural and human disturbance: flooding, tides, ice scouring, extremes in temperature, sediment accumulation, build up of debris; fluxes of nutrients, pollutants or sewage, and changes in hydroperiod (Dickinson and Miller 1998; Zedler and Kercher 2004).

Invasive plants find opportunities for potential establishment in the conditions created by these disturbances (Davis et al. 2000). The impacts of these invaders upon the native plant communities and the vulnerable wetlands they inhabit (Houlahan and Findlay 2004; Zedler and Kercher 2004) are not consistently studied in a quantitative manner. Of all the communities of organisms in freshwater tidal wetlands, the native plant community has the potential to be affected the most by invasive plants because both the community and the invader have similar resource requirements. Invasive plants co-opt space (Hager and McCoy 1998), and compete for light (Weihe and Neely 1998) and nutrients (Pfeifer-Meister et al. 2008, Sala et al. 2007). In addition, several studies have found that invasive plants have allelopathic effects on native plants (Nakai et al. 1999; Bais et al. 2003; Gross 2003; Hierro and Callaway 2003; Callaway et al. 2005; Morgan and Overholt 2005; Hilt 2006; Thelen et al. 2005; Rudrappa et al. 2007). Invasive-native plant interactions also present the possibilities of hybridization, parasitism, and the transfer of pathogens (Baker 1986; Simberloff 2003). Factors that affect the perceived outcome of each invasion are, in part, revealed through published studies of lab and field experiments. I review existing literature on the ecological effects of invasive wetland plants on native ecosystems and plant communities. This chapter discusses several common findings of those studies.

Methods

To access published studies, I created a list of invasive plants that are found in freshwater wetlands in North America based on information provided by the USDA Natural Resource Conservation Service website (http://plants.usda.gov) and by searching Web of Science using the search terms "invasive plants," introduced plants," "exotic plants," in combination with "wetlands." I chose to review studies published after 1995 on wetland plant species that had the largest potential ranges in North America wetlands and were of verified non-native origin, which produced a list of 15 species (Appendix A). I chose studies by searching Web of Science using scientific names or common names of the 15 species in combination with the search terms "species diversity," species richness," and "native plants." I previewed over one hundred records for the selected species and reviewed the papers (27) that had abstracts which mentioned that part or all of the study's focus was on the effects of invasive plants on native plants. In the case of several species, I was unable to find studies of their impacts on native plant communities in North America or in greenhouse experiments.

Results

The review of these publications revealed several important aspects of invasive plant-native plant interactions. Effects and factors are often difficult to distinguish from one another or may be quite subtle. However, I consistently encountered several variables that were associated with invasion impacts and outcomes in freshwater tidal wetlands. I have grouped these variables into two categories: experimental design and species-specific characteristics.

Experimental Design

Context

The context of study sites influences the ability of an invasive plant to become established or dominant. Land-use in neighboring areas can create disturbances (Waldner 2008; Gassó et al. 2009; Milbau et al. 2009), and impact nutrient availability and habitat vulnerability to invasion. Current and historical land-use of invaded areas and their surroundings can also affect the ability of invasive plants to become established (Hulme 2009; Mack et al. 2000; Waldner 2008; Wania et al. 2006). McGlynn (2006) found that human activity (proximity to an urban area) could have reduced or altered the native plant species composition of several wetland sites and affected the invasibility of wetlands surveyed. The urban wetlands studied by Ehrenfeld (2008) showed a decreased rate of plant invasions as industrialization in adjacent areas increased. Upland studies also confirm the importance of land-use in invasion outcomes (Endress et al. 2007; Aboh et al. 2008).

Seasonal Effects

Invader dominance seems to have a strong seasonal effect on native plants or its effects are modulated by seasonal effects that are related to light, nutrient availability, and possibly other abiotic factors. An invasive plant that arrives in the latter part of the growing season when many native plants are already adults might be unable to become established. Yakimowski et al. (2005) found that invasive *Lythrum salicaria* seedlings encountered limited light in *Typha*-dominated marshes (adult *Typha* spp. and plant litter) and their establishment was negatively affected. Unlike *Typha* spp., invasive *Crassula helmsii* inhibits growth of other plants at an even earlier stage—that of germination (Langdon et al. 2004). The growth rate of both native and invasive plants that are already established can also influence interactions. A greenhouse study performed by Mony et al. (2007) showed that invasive *Egeria densa* did not alter its biomass production and reproductive allocations as the seasons changed, but *Hydrilla verticillata* did and was able to outcompete *E. densa* for nutrients. Presumably *H. verticillata* outcompetes a number of native plants as well.

Temporal Scale

The length of time during which a study is conducted may determine the outcome observed. Over time the effects of an invasive plant may change. However, few of the studies I found occurred over a period of more than one or two seasons and all such studies focused on *L. salicaria*. Denoth and

Meyers (2007) removed the invasive *L. salicaria* from their study plots in the first year and found a significant increase in the vegetative performance of the rare native plant *Sidalcea hendersonii* in the first year, but not in the second year. In a three year study, Farnsworth and Meyerson (1999) found that native species richness and abundance increased the first year after removal of *P. australis*, but in the remaining years *Typha* sp., and then *P. australis*, became dominant. Mal et al. (1997) found that *L. salicaria* took four years to attain dominance in the study wetland. A microcosm experiment involving 20 plant species was dominated by *L. salicaria* in its fifth year (Weiher et al. 1996). In grasslands, native plant species exhibit a pattern in which their establishment is first dependent upon disturbance and then upon nutrients available for accumulation of biomass (Thompson et al. 2001); invasive plants in wetlands may follow a similar pattern.

Spatial Scale

Like temporal scale, landscape scale could also affect the perceived outcome of an invasion (Brown and Peet 2003; Kennedy et al. 2002; Milbau et al. 2009). Small scale studies found a negative association between native and invasive diversity with some researchers surmising that high native diversity decreases invasibility (Tilman 1997; Stachowicz et al. 1999; Levine 2000; Naeem et al. 2000; Lyons and Schwartz 2001; Fridley et al. 2007). Studies performed on large scales found positive relationships between native and invasive diversity (Lonsdale 1999). Davies et al. 2005 attributed this pattern to the spatial heterogeneity in the large landscapes (including heterogeneity of abiotic factors) that allowed for spatial heterogeneity in species composition while Shea and Chesson (2002) and Levine (2000) hypothesized that high native species diversity may also create more niche opportunities for invasive plants. Whatever the mechanisms behind the relationship between native and invasive diversity may be, studies in freshwater tidal wetlands occurred at both spatial scales and exhibited the predicted patterns in species richness and diversity. Hager and Vinebrook (2004) found a positive relationship between *L. salicaria* and native species diversity. Treberg and Husband (1999) found no evidence of a decrease in native plant diversity in wetlands invaded by *L. salicaria* and Massachusetts (United States) marshes dominated by both *L. salicaria* and native *Typha* spp. had high alpha diversity (Keller 2000). McGlynn (2006; 2009) found that species composition and diversity were significantly different among plots dominated by invasive *L. salicaria* or *P. australis* or native *Typha* spp. Native plant species richness decreased with increasing cover of *L. salicaria* (Gabor et al. 1996) and *L. salicaria* dominated the seed bank in experimental areas (Welling and Becker 1990). An important factor to consider before attributing an outcome to scale or species diversity is that the disturbance which created an opportunity for an invasion may have produced changes in the native plant community (McGlynn 2006; Gurevitch and Padilla 2004).

Metrics

Measures used to quantify invasive plant effects on native plants influence perceived outcomes of an invasion. Farnsworth and Ellis (2004) found that *L. salicaria* did not necessarily impact species richness and diversity, but did affect native plant biomass. Morrison (2002) also found that native plant cover and biomass was reduced in the presence of *L. salicaria*, but plant species richness did not change significantly. However, species richness and diversity were significantly affected by *Myriophyllum spicatum* even though there was no significant relationship between biomass of *M. spicatum* and native aquatic plants (Madsen et al. 2008). In two studies (Larsen 2007; Zhonghua et al. 2007) *Nymphoides peltata* negatively affected the growth (biomass) of several other species, *Elodea canadensis*, *Ranunculus circinatus*, and *Trapa spinosa*; but no species was eliminated. It may be that, as in other habitats, invasive plants in freshwater tidal wetlands do not generally cause native plants to become extinct (resulting in a change in species richness), but they can be associated with decreased abundance of a particular native species (possibly resulting in a change in species diversity) (Brown et al. 2006).

Species-Specific Characteristics

Zonation

The different vegetation zones in freshwater tidal marshes have different flooding regimes, elevations, and soils as well as different resident native plant species. An invader to a particular zone would have to possess particular characteristics in order to become established there (McGlynn 2006; 2009). For example, the ability of *Phragmites australis*, an invader to both brackish and freshwater wetlands, to become established was reduced as marsh elevation decreased and flooding frequency increased (Silliman and Bertness 2004). In fact, many of the species in the native plant communities that were invaded may have had not only similar tolerances for flooding, but also similar requirements for soil composition (Silliman and Bertness 2004). Farnsworth and Meyerson (2003) found that some species, e.g. *Leersia oryzoides*, had very specialized microenvironment requirements, while the dominant *Typha* sp. and *P. australis* did not. Thus the effect of zonation on plant invasions is very closely tied to species-specific characteristics.

Form and Function

Species-specific traits can determine the nature of interactions between invasive plants and the native plant communities they invade (Alvarez and Cushman 2002; Kourtev et al. 2003). They can affect how some plants' characteristics change when the plants become invasive (Willis et al. 2000), how closely the characteristics of an invader match those of native plants already present (Strauss et al. 2006), and how an invader affects an invaded community (Mason and French 2008). Some characteristics that are important in determining the outcome of native-invasive plant interactions include functional group (Mahaney et al. 2006), growth form (Hager 2004), and allelopathic ability (Gross 2003). If an invasive species becomes established in an area where it is the only species with a particular growth form or from a particular functional group it is likely to produce a more easily quantifiable effect on the native plant community (Vitousek et al. 1987; D'Antonio and Vitousek 1992; Vitousek 1992; Gordon 1998; Kourtev et al. 2003; Lesica and DeLuca 2004; Mahaney et al. 2006; Windham and Meyerson 2003). The heights of *L. salicaria* and *P. australis* enable them to limit the light available to other plant's seedlings (Odum et al. 1984; Rawinski and Malecki 1984; Gaudet and Keddy 1988; Edwards et al. 1995; Minchinton and Bertness 2003), while *H. verticillata* is able to shade and reduce root growth of fellow invasive *Myriophyllum spicatum* (Wang et al. 2008)

Dominance

Houlahan and Findlay (2004) concluded from their study of wetland plant communities in Southeastern Ontario, Canada that dominant invasive plants do not impact native plant diversity and species richness any more than native dominant plants. McGlynn (2006) also found that plots invaded by *L. salicaria* had the highest species richness relative to plots with the more dominant native *Typha* spp. or invasive *P. australis*, although the differences were not statistically significant. Morrison (2002) found that in plots where *L. salicaria* was dominant it did not form monospecific stands. When *L. salicaria* was removed from plots, several native species became dominant (Morrison 2002).

Conclusion

Many factors influence the outcomes of invasions and these outcomes are variable. A number of these factors are part of an experiment's design and the researcher can choose which of these factors will influence the outcome. Assuming that each invasion follows a predestined course with a negative outcome is not an effective approach for understanding the ecology of individual invasions or for making informed management decisions. Such assumptions also prevent ecologists from making use of the opportunity to study basic questions about community and ecosystem ecology that invasions provide.

Invasion science and management require empirical research that is conducted in both field and greenhouse, thoroughly quantifies actual ecological effects of each invasive plant species; and incorporates different measures of effect, multiple temporal scales, changes in the physical environment (i.e. zonation or changes in elevation that are characteristic of many tidal wetlands) and knowledge of the importance of species characteristics of all plants (both native and invasive) involved in the interactions.

Even though *Alternanthera philoxeroides*, *Butomus umbellatus*, *Eichhornia crassipes*, *Hydrocharis morsus-ranae*, *Najas minor*, *Panicum repens*, *Salvinia molesta*, and *Trapa natans* have extensive potential ranges for invasion in North America, surprisingly few studies have been published about invasions involving these floating and submerged species. This may be due to the difficulty of conducting experiments in tidal open water habitats. Much information is needed about these species' interactions with native plants, native vertebrates and invertebrates, as well as their potential ecosystem effects. Once this information has been collected it should be made widely available so that decisions can be made about whether or not a particular invasive plant is an imminent threat and what can be done about it. Invasion biology and its associated management needs have evolved beyond anecdotal data.

Acknowledgment

My thanks to Dr. Kerry Brown, Dr. Elizabeth Farnsworth, Dr. Eliza Woo, and Dr. Helen Bustamante Wood for their comments on this chapter.

Appendix A. Invasive species with Largest Potential Wetland Ranges in North America and of Verified Non-Native Origin

Scientific Name	Location
Alternanthera philoxeroides (Mart.) Griseb.	15 U.S. states
Butomus umbellatus L.	16 U.S. states and 8 Canadian provinces
Crassula helmsii A. Berger	At least 3 U.S states
Eichhornia crassipes (Mart.) Solms	24 U.S. states and 1 Canadian province
Egeria densa Planch.	37 U.S. states and 1 Canadian province
Hydrilla verticillata (L.f.) Royle	20 U.S. states
Hydrocharis morsus-ranae L.	1 U.S. states and 2 Canadian provinces
Lythrum salicaria L.	44 U.S. states and 10 Canadian provinces
Myriophyllum spicatum L.	43 U.S. states and 4 Canadian provinces
Najas minor All.	26 U.S. states and 1 Canadian province
Nymphoides peltata (S.G. Gmel.) Kuntze	25 U.S. states and 2 Canadian provinces
Panicum repens L.	10 U.S. states
Phragmites australis (Cav.) Trin. ex. Steud. Invasive status (haplotype) not verified in each state or province	49 U.S. states, Puerto Rico and 11 Canadian provinces
Salvinia molesta Mitchell	11 U.S. states
Trapa natans L.	9 U.S. states and 1 Canadian province

Appendix B. The Papers Reviewed

Citation	Invasive species
Denoth and Meyers 2007	*Lythrum salicaria*
Farnsworth and Ellis 2001	*Lythrum salicaria*
Farnsworth and Meyerson 1999	*Phragmites australis*
Farnsworth and Meyerson 2003	*Phragmites australis*
Gabor et al. 1996	*Lythrum salicaria*
Hager 2004	*Lythrum salicaria*
Hager and Vinebrook 2004	*Lythrum salicaria*
He et al. 2008	*Najas minor*
Houlahan and Findlay 2004	*Lythrum salicaria*
Keller 2000	*Lythrum salicaria* and *Phragmites australis*
Langdon et al. 2004	*Cressula helmsii*
Larson 2007	*Nymphoides peltata*
Madsen et al. 2008	*Myriophyllum spicatum*
Mahaney et al. 2006	*Lythrum salicaria*
Mal et al. 1997	*Lythrum salicaria*
McGlynn 2009	*Lythrum salicaria* and *Phragmites australis*
Mony et al. 2007	*Egeria densa*
Morrison 2002	*Lythrum salicaria*
Rawinski and Malecki	*Lythrum salicaria*
Treberg and Husband 1999	*Lythrum salicaria*
Wang et al. 2008	*H. verticillata*
Weihe and Neely 1998	*Lythrum salicaria*
Weiher et al. 1996	*Lythrum salicaria*
Welling and Becker 1990	*Lythrum salicaria*
Yakimowksi et al. 2005	*Lythrum salicaria*
Zhonghua et al. 2007	*Nymphoides peltata*

References

[1] Aboh, B. A., Hovinato, M., Oumorou, M. and Sinsin. B. 2008. Invasiveness of two exotic species, *Chromolaena odorata* (Asteraceae)

and *Hyptis suaveolens* (Lamiaceae), in relation with land use around Betecoucou (Benin). *Belgian Journal of Botany* 141(2): 125-140.

[2] Alvarez, M. E. and Cushman, J. H. 2002. Community level consequences of a plant invasion: effects on three habitats in coastal California. *Ecological Applications* 12(5): 1434-1444.

[3] Bais, H. P., Vepachedu, R., Gilroy, S., Callaway, R. M. and Vivanco, J. M. 2003. Allelopathy and exotic plant invasions: From molecules and genes to species interactions. *Science* 301(5638): 1377-1380.

[4] Baker, H. G. 1986. Patterns of plant invasions in North America. In: Mooney, H. A. and Drake, J. A., editors. *Ecology of Biological Invasions of North America and Hawaii,* New York: Springer-Verlag. 44-57.

[5] Brown, R. L. and Peet. R. K. 2003. Diversity and invasibility of southern Appalachian plant communities. *Ecology* 84(1): 32-39.

[6] Brown, K., Scatena, F. N. and Gurevitch, J.2006. Effects of an invasive tree on community structure and diversity in a tropical forest in Puerto Rico. *Forest Ecology and Management.* 226(1-3): 145-152.

[7] Callaway, R. M., Ridenour, W. M., Laboski, T., Weir, T. and Vivanco, J. M. 2005. Natural selection for resistance to the allelopathic effects of invasive plants. *Journal of Ecology* 93(3): 576-583.

[8] Carpenter, S. R. and Lodge. D. M.1986. Effects of submersed macrophytes on ecosystem processes. *Aquatic Botany* 26: 341-370.

[9] D'Antonio, C. M. and Vitousek, P. M. 1992. Biological invasions by exotic grasses, the grass/fire cycle and global change. *Annual Review of Ecology and Systematics*. 23: 63-87.

[10] Davies, K., Chesson, P., Harrison, S., Inouye, B. D., Melbourne, B. A. and Rice, K. J. 2005. Spatial heterogeneity explains the scale dependence of native-exotic diversity relationship. *Ecology* 86 (6): 1602-1610.

[11] Davis, M. A., Grime, J. P. and Thompson. K. 2000. Fluctuating resources in plant communities: a general theory of invasibility. *Journal of Ecology* 88:528-534.

[12] Denoth, M. and Meyers, J. H. 2007. Competition between *Lythrum salicaria* and a rare species: combining evidence from experiments and long-term monitoring. *Plant Ecology* 191: 153-161.

[13] Dickinson, M. B. and Miller, T. E. 1998. Competition among small, free-floating, aquatic plants. *American Midland Naturalist* 140 (1): 55-67.

[14] Edwards, K. R., Adams, M. S. and Květ, J. 1995. Invasion history and ecology of *Lythrum salicaria* in North America. In: *Plant Invasions: General Aspects and Special Problems,* eds. P. Pyšek, K. Prach, M. Rejmánek and M. Wade. Pp 39-60. SPB Academic Publishing, Netherlands.

[15] Ehrenfeld, J. G. 2008. Exotic invasive species in urban wetlands: environmental correlates and implications for environmental management. *Journal of Applied Ecology* 45: 1160-1169.

[16] Endress, B. A., Naylor, B. J., Park, G. G. and Radosevich, S. R. 2007. Landscape factors influencing the abundance and dominance of the invasive plant *Potentilla recta. Rangeland Ecology and Management.* 60(3): 218-224.

[17] Farnsworth, E. J. and Ellis, D. R. 2001. Is purple loosestrife (*Lythrum salicaria*) an invasive threat to freshwater wetlands? Conflicting evidence from several ecological metrics. *Wetlands* 21(2): 199-209.

[18] Farnsworth, E. J. and Meyerson, L. A. 1999. Species composition and inter-annual dynamics of freshwater tidal plant community following removal of the invasive grass, *Phragmites australis. Biological Invasions* 1:115-127.

[19] Farnsworth, E. J. and Meyerson. L. A. 2003. Comparative ecophysiology of four wetland plant species along a continuum of invasiveness. Wetlands 23(4): 750-762.

[20] Fridley, J. D., Stachowicz, J. J., Naeem, S., Sax, D. F., Seabloom, E. W., Smith, M. D., Stohlgren, T. J., Tilman, D. and Von Holle, B. 2007. The invasion paradox: Reconciling pattern and process in species invasions. *Ecology* 88(1): 3-17.

[21] Gabor, T. S., Haagsma, T. and Murkin, H. R. 1996. *Wetland plant responses to varying degrees of purple loosestrife removal in Southeastern Ontario*, Canada. Wetlands 16:95-98.

[22] Gassó, N., Sol, D., Pino, J., Dana, E. D., Lloret, F., Sanz-Elorza, M., Sobrino, E. and Vilà. M. 2009. Exploring species attributes and site characteristics to assess plant invasions in Spain. *Diversity and Distributions* 15:50-58.

[23] Gaudet, C. L. and Keddy, P. A. 1989. A comparative approach to predicting competitive ability from plant traits. *Nature* 334(6169): 242-243.

[24] Gordon, D. R. 1998. *Effects of invasive, non-indigenous plant species on ecosystem processes: Lessons from Florida. Ecological Applications* 8(4): 975-989.

[25] Gross, E. 2003. Allelopathy of aquatic autotrophs. *Critical Reviews in Plant Science* 22: 313-339.

[26] Gurevitch, J. and Padilla, D. 2004. Are invasive species a major cause of extinctions? *Trends in Ecology and Evolution*. 19(9): 470-474.

[27] Hager, H. A. 2004. Competitive effect versus competitive response of invasive and native wetland plant species. *Oecologia* 139:140-149.

[28] Hager, H. A. and McCoy, K. D. 1998. The implications of accepting untested hypotheses: a review of the effects of purple loosestrife (*Lythrum salicaria*) in North America. *Biodiversity and Conservation* 7(8): 1069-1079.

[29] Hager, H. A. and Vinebrook. R. D. 2004. Positive relationships between invasive purple loosestrife (*Lythrum salicaria*) and plant species diversity and abundance in Minnesota wetlands. *Canadian Journal of Botany* 82(6): 763-773.

[30] He, F., Deng, P., Wu, X. H., Cheng, S. P., Gao, Y. N. and Wu. Z. B. 2008. Allelopathic effects on *Scenedesmus obliquus* by two submerged macrophytes *Najas minor* and *Potamogeton malaianus*. *Fresenius Environmental Bulletin* 17(1): 92-97.

[31] Hierro, J. L. and Callaway, R. M. 2003. Allelopathy and exotic plant invasion. *Plant and Soil 256*(1): 29-39.

[32] Hilt, S. 2006. Allelopathic inhibition of epiphytes by submerged macrophytes. *Aquatic Botany* 85(3): 252-256.

[33] Houlahan, J. E. and Findlay, C. S. 2004. Effect of invasive plant species on temperate wetland plant diversity. *Conservation Biology* 18(4): 1132-1138.

[34] Hulme, P. E. 2009. Relative role of life-form, land use and climate in recent dynamics of alien plant distributions in the British Isles. *Weed Research* 49: 19-28.

[35] Keller, B. E. M. 2000. Plant diversity in *Lythrum*, *Phragmites* and *Typha* marshes, Massachusetts, U. S. A. *Wetlands Ecology and Management* 8: 391-401.

[36] Kennedy, T. A., Naeem, S., Howe, K. M., Knops, J. M. H., Tilman, D. and Reich, P. 2002. Biodiversity as a barrier to ecological invasion. *Nature* 417: 636-638.

[37] Kourtev, P. S., Ehrenfeld, J. G. and Haggblom. M. 2003. Experimental analysis of the effect of exotic and native plant species on the structure and function of soil microbial communities. *Soil Biology and Biochemistry* 35(7): 895-905.

[38] Langdon, S. J., Marrs, R. H., Hosie, C. A., H. A., K. M. Norris and Potter, J. A. 2004. *Crassula helmsii* in UK ponds: Effects on plant biodiversity and implications for newt conservation. *Weed Technology* 18: 1349-1352.

[39] Larsen, D. 2007. Growth of three submerged plants below different densities of *Nymphoides peltata* (S. G. Gmel.) Kuntze. *Aquatic Botany* 86: 280-284.

[40] Lesica, P. and DeLuca, T. H. 2004. Is tamarisk allelopathic? *Plant and Soil* 267(1-2): 357-365.

[41] Levine, J. M. 2000. Species diversity and biological invasions: Relating local process to community pattern. *Science* 288(5467): 852-854.

[42] Lonsdale, W. M. 1999. Global patterns of plant invasions and the concept of invasibility. *Ecology* 80:1522-1536.

[43] Lyons, K. G. and Schwartz, M. W. 2001. Rare species alters ecosystem function – invasion resistance. *Ecology Letters* 4: 358-365.

[44] Mack, R. N., Simberloff, D., Lonsdale, W. M., Evans, H., Clout, M.and Bazzazz, F. 2000. Biotic Invasions: Causes, epidemiology, global consequences and control. *Issues in Ecology Number* 5. 20 pp.

[45] Madsen, J. D., Stewart, R. M., Getsinger, K. D., Johnson R. L., and Wersal. R. M. 2008. Aquatic plant communities in Waneta Lake and Lamoka Lake, New York. *Northeastern Naturalist* 15(1): 97-110.

[46] Mahaney, W. M., Smemo, K. A. and Yavitt, J. B. 2006. Impacts of *Lythrum salicaria* on plant community and soil properties in two wetlands in central New York, USA. *Canadian Journal of Botany* 84: 477-484.

[47] Mal, T. K., Lovett-Doust, J. and Lovett-Doust, L. 1997. Time-dependent competitive displacement of *Typha angustifolia* by *Lythrum salicaria*. *Oikos* 79:26-33.

[48] Mason, T. J. and French. K. 2008. Impacts of a woody invader vary in different vegetation communities. *Diversity and Distributions* 14: 829-838.

[49] McGlynn, C. A. 2006. The effects of two invasive plants on native communities in Hudson River freshwater tidal wetlands. *Dissertation.* State University of New York, Stony Brook, NY. 218 pp.

[50] McGlynn, C. A. 2009. Native and invasive plant interactions in wetlands and the minimal role of invasiveness. *Biological Invasions*. (In press)

[51] Minchinton, T. E. and Bertness, M. D. 2003. Disturbance-mediated competition and the spread of *Phragmites australis* in a coastal marsh. *Ecological Applications* 13(5): 1400-1416.

[52] Milbau, A., Stout, J. C., Graae, B. J. and Nijs, I. 2009. A hierarchical framework for integrating invasibility experiments incorporating different factors and spatial scales. *Biological Invasions* 11: 941-950.

[53] Mony, C., Koschnick, T. J., Haller, W. T. and Muller, S. 2007. Competition between two invasive Hydrocharitaceae (*Hydrilla verticillata* (L. f) (Royle) and *Egeria densa* (Planch) as influenced by sediment fertility and season. *Aquatic Botany* 86: 236-242.

[54] Morgan, E. C. and Overholt, W. A. 2005. Potential allelopathic effects of Brazilian pepper (*Schinus terebinthifolius* Raddi, Anacardiaceae) aqueous extract on germination and growth of selected Florida native plants. *Journal of Torrey Botanical Society* 132(1): 11-15.

[55] Morrison, J. A. 2002. Wetland vegetation before and after experimental purple loosestrife removal. *Wetlands* 22(1): 159-169.

[56] Naeem, S., J., Knops, M. H., Tilman, D., Howe, K. M., Kennedy, T. and Gale, S. 2000. Plant diversity increases resistance to invasion in the absence of covarying extrinsic factors. *Oikos* 91: 97-108.

[57] Odum, W. E., Smith, III, T. J., Hoover, J. K. and McIvor, C. C. 1984. The ecology of tidal freshwater marshes of the United States East Coast: A community profile. *U. S. Fish and Wildlife Service*, FWS/OBS-83/17.

[58] Pfeifer-Meister, L., Cole, E. M., Roy, B. A. and Bridgham, S. D. 2008. Abiotic constraints on the competitive ability of exotic and native grasses in a Pacific Northwest prairie. *Oecologia* 155(2): 357-366.

[59] Rawinski, T. J. and Malecki, R. A. 1984. Ecological relationships among purple loosestrife, cattail and wildlife at the Montezuma National Wildlife Refuge. *New York Fish and Game Journal* 31:81-87.

[60] Rudrappa, T., Bonsall, J., Gallagher, J. L., Seliskar, D. M. and Bais, H. P. 2007. Root-secreted allelochemical in the noxious weed *Phragmites australis* deploys a reactive oxygen species response and microtubule assembly disruption to execute rhizotoxicity. *Journal of Chemical Ecology* 33(10): 1573-1561.

[61] Sala, A., Verdaguer, D. and Vila, M. 2007. Sensitivity of the invasive geophyte *Oxalis pes-caprae* to nutrient availability and competition. *Annals of Botany* 99(4): 637-645.

[62] Shea, K. and Chesson. P. 2002. Community ecology theory as a framework for biological invasions. *Trends in Ecology and Evolution* 17(4): 17-176.

[63] Silliman, B. R. and Bertness, M. D. 2004. Shoreline development drives invasion of *Phragmies australis* and the loss of plant diversity on New England salt marshes. *Conservation Biology* 18(5):1424-1434.
[64] Simberloff, D. 2003. Confronting introduced species: a form of xenophobia? *Biological Invasions* 5:179-192.
[65] Stachowicz, J. J., Whitlatch, R. B. and. Osman, R. W 1999. Species diversity and invasion resistance in a marine ecosystem. *Science* 286: 1577-1579.
[66] Strauss, S. Y., Webb, C. O. and Salamin, N. 2006. *Exotic taxa less related to native species are more invasive*. Proceedings of the National Academy of Science 103(15): 5841-5845.
[67] Tilman, D. 1997. Community invasibility, recruitment limitation and grassland biodiversity. *Ecology* 78(1): 81-92.
[68] Treberg, M. A. and Husband, B. C. 1999. Relationship between the abundance of *Lythrum salicaria* (purple loosestrife) and plant species richness along the Bar River, Canada. *Wetlands.* 19(1): 118-125.
[69] Thelen, G. C., Vivanco, J. M., Newingham, B., Good, W., Bais, H. P., Landres, P. Ceasar, A. and Callaway, R. M. 2005. Insect herbivory stimulates allelopathic exudation by an invasive plant and the suppression of natives. *Ecology Letters* 8(2): 209-217.
[70] Thompson, K., Hodgson, J. G., Grime, J. P. and Burke, M. J. W. 2001. Plant traits and temporal scale: evidence from 5-year invasion experiment using native species. *Journal of Ecology* 89:1054-1060.
[71] Vitousek, P. M., Walker, L. R., Whittaker, L. D., Mueller-Dombois, D. and Matson, P. A. 1987. Biological invasion by *Myrica faya* alters ecosystem development in Hawaii. Science 238: 802-804.
[72] Vitousek, P. M. 1992. Effects of alien plants on native ecosystems. In: Stone, C. P., Smith, C. W. and Tunison, J. T., editors. Alien plant invasions in native ecosystems of Hawai'i. University of Hawai'i Cooperative National Park Resources Studies Unit, 3190 Maile Way, Honolulu. 29-41.
[73] Waldner, L. S. 2008. The kudzu connection: Exploring the link between land use and invasive species. *Land Use Policy* 25: 399-409.
[74] Wang, J. W., Yu, D., Xiong, W. and Han, Y. Q. 2008. Above- and belowground competition between two submersed macrophytes. *Hydrobiologia* 607:113-122.
[75] Wania, A., Kühn, I. and Klotz. S. 2006. Plant richness patterns in agricultural and urban landscapes in Central Germany-spatial gradients of species richness. 2006. *Landscape and Urban Planning* 75: 97-110.

[76] Weihe, P. E. and Neely, R. K. 1998. The effects of shading on competition between purple loosestrife and broad-leaved cattail. *Aquatic Botany* 59(1-2): 127-138.
[77] Weiher, E., Wisheu, I. C., Keddy, P. A. and Moore, D. R. J. 1996. Establishment, persistence and management implications of experimental wetland plant communities. *Wetlands.* 16(2): 208-218.
[78] Welling, C. H. and Becker, R. L. 1990. Seed bank dynamics of *Lythrum salicaria* L. : Implications for control of this species in North America. *Aquatic Botany* 38: 303-309.
[79] Wilcove, D. S., Rothstein, D., Dubow, J., Phillips, A. and Losos, E. 1998. Quantifying threats to imperiled species in the United States. *BioScience* 48(8): 607-615.
[80] Willis, A. J., Memmott, J. and Forrester, R. I. 2000. Is there evidence for the post-invasion evolution of increased size among invasive plant species? *Ecological Letters* 3(4): 275-283.
[81] Windham, L. and Meyerson, L. A. 2003. Effects of common reed (*Phragmites australis*) expansions on nitrogen dynamics of tidal marshes of the Northeastern U. S. Estuaries 26(2B): 452-464.
[82] Yakimowski, S. B., Hager, H. A. and Eckert, C. G. 2005. Limits and effects of invasion by the nonindigenous wetland plant *Lythrum salicaria* (purple loosestrife): a seed bank analysis. *Biological Invasions* 7:687-698.
[83] Zedler, J. B. and Kercher, S. 2004. Causes and consequences of invasive plants in wetlands: Opportunities, Opportunists and Outcomes. *Critical Reviews in Plant Sciences* 23(5): 431-452.
[84] Zhonghua, W., Dan, Y., Manghui, T., Qiang, W. and Wen, X. 2007. Interference between two floating-leaved aquatic plants: *Nymphoides peltata* and *Trapa bispinosa*. Aquatic Botany 86: 316-320.

INDEX

C

D

E

F

G

H

I

J

K

L

M

Q

R

S

T

U

V

W

X

Y

Z